ATMOSPHERE and OCEAN

ATMOSPHERE and OCEAN

Our Fluid Environments

by

JOHN G. HARVEY

Senior Lecturer in Environmental Sciences at the University of East Anglia

1ST EDITION

PUBLISHED BY

THE ARTEMIS PRESS

SUSSEX MCMLXXVI

FIRST PUBLISHED 1976

ISBN 0 85141 296 3 (library edition)

ISBN 0 85141 295 5 (paperback edition)

Set in 11pt Bembo

Printed in Great Britain by
Balding & Mansell Ltd., Wisbech, Cambridgeshire.

CONTENTS

Preface 8

Acknowledgements 9

1 **Origins and Geological Background** 11

Evolution of the atmosphere and ocean water □ Origin and topography of ocean basins

2 **Fluids in General: The Atmosphere and Water in Particular** 15

Compressibility and hydrostatic pressure □ Composition of the atmosphere □ Adiabatic changes □ Viscosity □ The peculiar properties of water

3 **The Hydrological Cycle: Water in the Atmosphere** 21

Stages and processes of the hydrological cycle □ Evaporation □ Saturation □ Vapour pressure □ Dewpoint □ Relative humidity □ Condensation □ Fog □ Cloud formation □ Cloud types □ Growth of cloud droplets □ Precipitation

4 **Salts, Gases and Ice in the Oceans** 31

Constituents of sea water □ Salinity □ Nutrient salts □ Dissolved gases: oxygen and carbon dioxide □ Formation of sea-ice □ Icebergs

5 **The Energy Source and Heat Distribution** 39

Solar radiation □ Transmission through the atmosphere □ Effects of the Earth's surface □ Albedo □ Heat balance and back radiation □ Heat transfer within and between the atmosphere and ocean □ Diurnal and seasonal variations of temperature □ Global distribution of air and sea surface temperature

6 Vertical Stability and Temperature Distribution 51

Stability of a fluid □ Stability in the oceans and atmosphere □ Adiabatic changes and potential temperature □ Assessment of stability: σ_t in the ocean, adiabatic lapse rates in the atmosphere □ Föhn winds □ Vertical temperature structure of the atmosphere: troposphere (average lapse rates and inversions), stratosphere, mesosphere and thermosphere □ Vertical temperature structure of the ocean and lakes □ Thermoclines

7 Air Masses and Water Masses 59

The air mass concept □ Characteristics, source regions, modification □ Classification and associated weather, with examples from British Isles □ The extension of this concept to water masses □ T/S diagrams □ Conservative and non-conservative properties □ Examples of water masses, and their rôle in tracing sub-surface water movement in the oceans

8 Motions and Forces 65

Relationship between force and motion □ Steady state conditions □ Laminar and turbulent flow □ Scales of motion □ Types of forces involved □ Centrifugal force □ Coriolis force

9 Waves and Tides 73

Progressive waves: general characteristics, deep-water waves and long waves □ Wind waves: generation, main features, shallow water effects □ Tsunamis □ Standing waves: general characteristics □ Tide-generating forces □ The equilibrium tide □ Tidal phenomena in the ocean and in marginal seas: resonance, diurnal and semi-diurnal tides, amphidromic points, tidal currents, tide prediction

10 Pressure Gradients and Associated Winds 85

Land and sea breezes □ Horizontal pressure gradients and associated force □ Anabatic and katabatic winds □ Geostrophic, surface, gradient and cyclostrophic winds

11 Geostrophic Currents and Thermal Winds 93

Geostrophic currents and their relationship to the slope of the sea surface and to density distribution □ Thermal winds

12 Atmospheric Circulation 99

Pattern on a uniform, non-rotating globe □ Actual surface pressure and wind patterns □ Effect of Earth's rotation □ Dishpan experiments □ The Hadley cell □ The upper westerlies, jet streams and Rossby waves □ The mid-latitude westerlies □ Influence of continents □ Monsoons

13 **Cyclones and Anticyclones** 109

Mid-latitude frontal depressions □ Formation in relation to upper westerlies □ Life history and associated weather □ Tropical cyclones □ Formation, warning signs, associated weather □ Avoiding action for shipping □ Anticyclones: formation and associated weather

14 **Oceanic Circulation** 119

Frictional coupling between the atmosphere and ocean □ Ekman spiral and storm surges □ Circulation of upper layers of the ocean □ Movement of icebergs □ Convergence and divergence □ Upwelling □ The deep circulation

Summary of Mathematical Expressions 129

List of Further Reading 133

Index 137

PREFACE

THE PURPOSE OF THIS BOOK is to provide an integrated introduction to meteorology and physical oceanography. With experience of teaching each subject independently in first-year university courses, I have become convinced that an integrated course is very much more satisfactory. There are two main reasons for this. The first is that the same basic principles apply to many aspects of the two subjects, as both are concerned with the physical properties and movements of fluids at and near the surface of the rotating Earth. Thus concepts such as hydrostatic pressure, vertical stability, Coriolis effect, geostrophic flow and vorticity are common to both, and the learning process is consolidated and becomes more efficient when the different applications of a particular concept are brought together. The second reason is that the atmosphere and the ocean interact with one another to a very considerable extent. The exchanges of thermal energy, momentum and water across the interface between the sea and the air play major rôles in determining the physical characteristics of each fluid, and the exchanges of gases (in particular carbon dioxide and oxygen) and salts are also of considerable though more limited significance. A study of either the atmosphere or the ocean must therefore be making constant reference to the other, and to the movements and distributions of properties within it.

During recent years I have had the opportunity to develop such an integrated course for the first year of the Environmental Sciences degree programme at the University of East Anglia, and this book has essentially resulted from it. It will, I hope, prove suitable for introductory courses for oceanography, meteorology, geography and environmental science students in other universities, polytechnics, and colleges of education, and for sea-going cadets who are taking courses in nautical science. If it enables teachers without previous knowledge of both of our fluid environments to introduce their students to them in an integrated way, I shall be especially gratified.

Some rudimentary mathematical knowledge is assumed, in particular familiarity with SI units, vectors and the differential coefficient; but any reader should be able to interpret the meaning of the equations used in this book with the aid of the Summary of Mathematical Expressions on pp. 129–132.

I am indebted to all of my colleagues and students who have contributed directly or indirectly to the development of the course from which this book has grown, but I should particularly like to express my thanks to those of my colleagues who read early drafts of various sections of this book and made helpful comments on them – Professor K. Clayton, Dr. T. Davies, Professor H. Lamb, Dr. P. Liss, Dr. C. Vincent, and Professor F. Vine. I am also very grateful to Mr. D. Mew who prepared all of the drawings for publication; to my wife Chris who read and corrected the manuscript many times and gave me constant encouragement; and to Mr. M. Bizony of Artemis Press, who made many helpful alterations to the text in the interests of clarity. I should also like to acknowledge the speed and efficiency with which Artemis Press have handled the publication of this book.

This book is concerned mainly with the physical aspects of the atmosphere and the ocean. For completeness, brief accounts are given in the early chapters of the nature and origin of the ocean basins and of the chemical composition of the atmosphere and sea water, but the reader who is interested in pursuing these topics further should refer to other books. No attempt has been made to cover the biological aspects, or man's interrelationships with the atmosphere and ocean. The economic justification for resources being devoted to the study of the atmosphere and ocean is well made in various books devoted specifically to the consideration of the direct and indirect ways in which man uses and is influenced by his fluid environments, as well as the impact he can have on them. Appreciation of the economic benefits to be gained has led to the establishment of numerous national and international organisations to foster and carry out research.

J. G. HARVEY

School of Environmental Sciences,
University of East Anglia,
Norwich.
May 1976.

Acknowledgements:

A number of illustrations and some statistics were directly based on, or loosely inspired by, material first published by the following authors and organizations, to whom we hereby make grateful acknowledgement on behalf of the author and ourselves:

R. G. Barry, R. J. Chorley and Methuen & Co., Ltd.; Heinrich Berann and Alcoa Ltd.; T. Bergeron; A. Defant; G. Dietrich and John Wiley & Sons, Inc.; D. Fultz; P. Groen; M. G. Gross; J. C. Johnson; P. S. Liss; D. H. McIntosh and A. S. Thom; J. Namias and P. F. Clapp; S. Petterssen, H. Riehl and McGraw Hill & Co.; W. J. Pierson et al.; H. Stommel; A. N. Strahler and Harper Row Publications, Inc.; H. U. Sverdrup, Martin W. Johnson, Richard H. Fleming and Prentice Hall, Inc.; J. Williams, J. Higginson and J. Rohrbough; G. Wuest.

Aerofilms, Ltd.; the Controller, H. M. Stationery Office; the Hydrographer of the Royal Navy; the Meteorological Office; N.A.S.A. and the United States Embassy, London; the National Institute of Oceanographic Sciences; the Royal Geographical Society; the Scott Polar Institute, University of Cambridge; Seaphot; the Smithsonian Institution; the United States Naval Institute, Annapolis; the United States Oceanographic Office.

We also express our thanks to Mr. Robert Hatch and his colleagues at our printers, Messrs. Balding & Mansell, for the careful and intelligent attention they gave this book in the press.

ARTEMIS PRESS

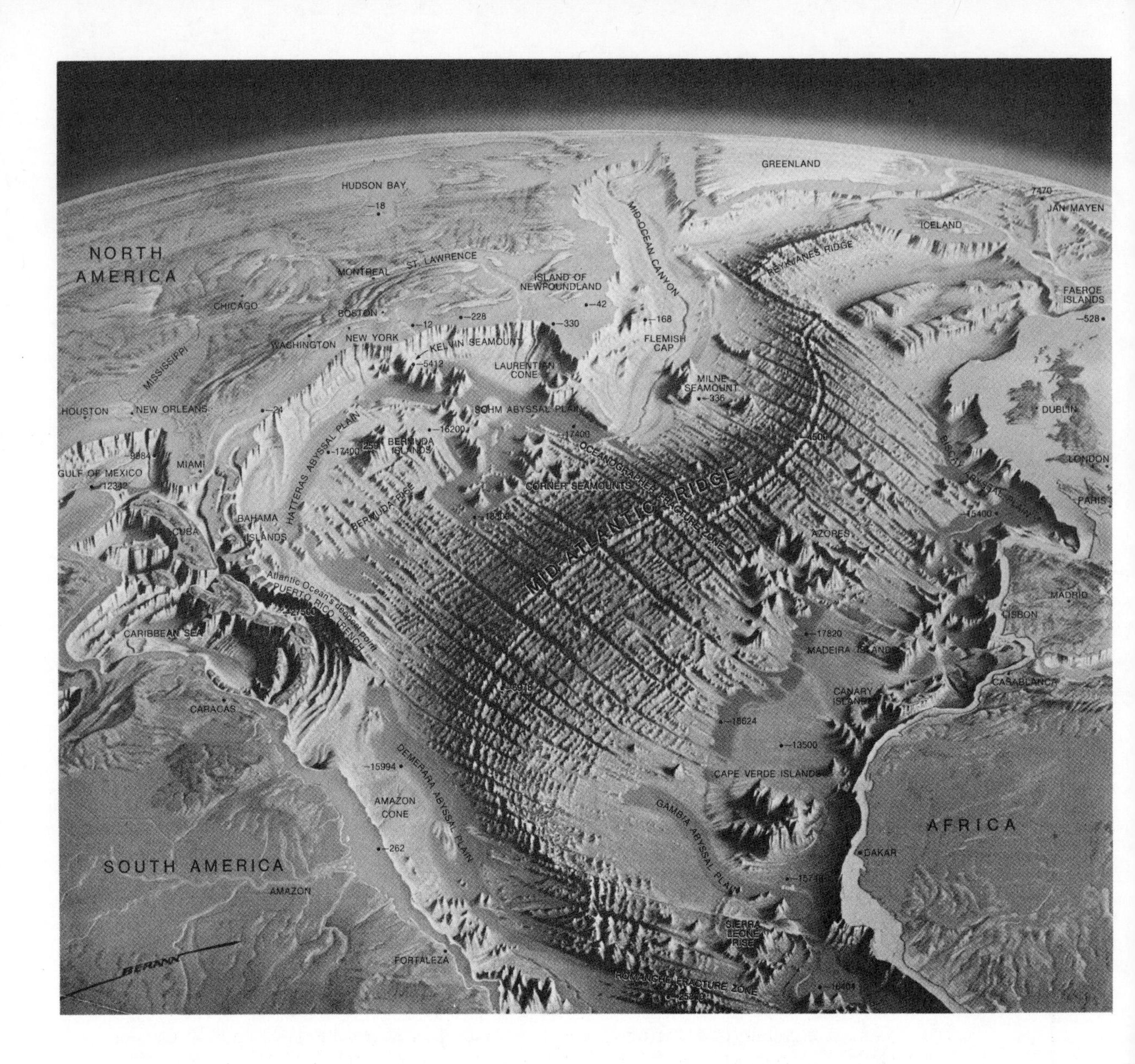

1.0 An artist's impression of the ocean floor topography of the North Atlantic basin. The sinuous, faulted structure running diagonally across the middle is the Mid-Atlantic Ridge, which like other ocean ridges is the result of sea-floor spreading. (From a painting by Heinrich C. Berann, based on bathymetric studies by Bruce C. Heezen and Marie Tharp of the Lamont Geological Observatory.)

1 Origins and Geological Background

Evolution of the atmosphere and ocean water □ Origin and topography of ocean basins.

OUR ATMOSPHERE has evolved over geological time, and the development of life on the Earth has been closely related to the composition of the atmosphere. Any primary atmosphere which the Earth had when it was formed some 4 600 million years ago appears to have been driven off during a period of heating, and the present atmosphere is a secondary one with constituents which have come from within the Earth. Gases given off during volcanic activity in the Earth's history have probably included hydrogen (H_2), water vapour (H_2O), carbon monoxide (CO) and dioxide (CO_2), nitrogen (N_2), hydrogen sulphide (H_2S) and hydrogen chloride (HCl). The relative proportions in which these were emitted changed as the interior of the Earth changed, in particular as the core and the mantle separated. Some theories suggest that most of the Earth's degassing occurred in a relatively short period in the Earth's history, while others postulate that it has continued steadily over most of geological time. The gases, once emitted, were subject to dissociation by sunlight, and reacted together. Thus methane (CH_4) and ammonia (NH_3) were formed, water vapour condensed and carbon dioxide, hydrogen chloride and ammonia dissolved in it.

From the geological record it seems that it was about 1 500 million years ago that free oxygen first appeared in the atmosphere in appreciable quantities. Before that the oxygen produced from the photo-dissociation of water vapour was fully utilised in the oxidation of surface materials. The evolution of life was very dependent on the availability of oxygen, but once sufficient had accumulated for green plants to develop, photosynthesis was able to liberate more into the atmosphere. The present concentration of oxygen, which appears to represent a state of dynamic equilibrium between its production and consumption, was probably reached about 100 or 200 million years ago.

Of the early atmospheric constituents, carbon comprises a major part of the material laid down in sediments as carbonates (e.g. limestone, $CaCO_3$) and fossil fuels (e.g. coal and oil); the water has formed the oceans and has considerable amounts of chlorides dissolved in it. Other chlorides are found in salt deposits (e.g. $NaCl$), and sulphur occurs in sedimentary rocks (e.g. as iron pyrites, FeS_2). Nitrogen has accumulated mainly in the atmosphere, whereas hydrogen has been lost to space because of its low molecular weight and because the high temperatures in the upper atmosphere give hydrogen molecules there velocities great enough to escape the Earth's gravitational field. Argon and helium have been added to the atmosphere by the radioactive breakdown of potassium, uranium and thorium, but whereas argon has accumulated in the atmosphere, helium, like hydrogen, has a sufficiently low molecular weight to escape to space.

So *water* has existed on the Earth's surface since fairly early in geological history, and was the solvent for substances derived from the Earth's crust by weathering, as well as for some of those emitted by volcanism and dissolved from the atmosphere. These latter, which have come from the degassing of the Earth's interior rather than by condensation from the primitive nebula when the planet was formed, have been called the 'excess volatiles', and include chlorine, bromine, sulphur and boron in addition to carbon, nitrogen and of course water itself. Geological evidence from sedimentation supports the view that there was a considerable amount of water on the Earth's surface at least as far back as 3 000 million years ago.

The present *ocean basins*, however, are relatively young features of the Earth, almost all less than 250 million years old (a mere 5% of geological time). We must here distinguish the oceanic crust, which is mainly between 2 km and 6 km below present sea level, from the older continental

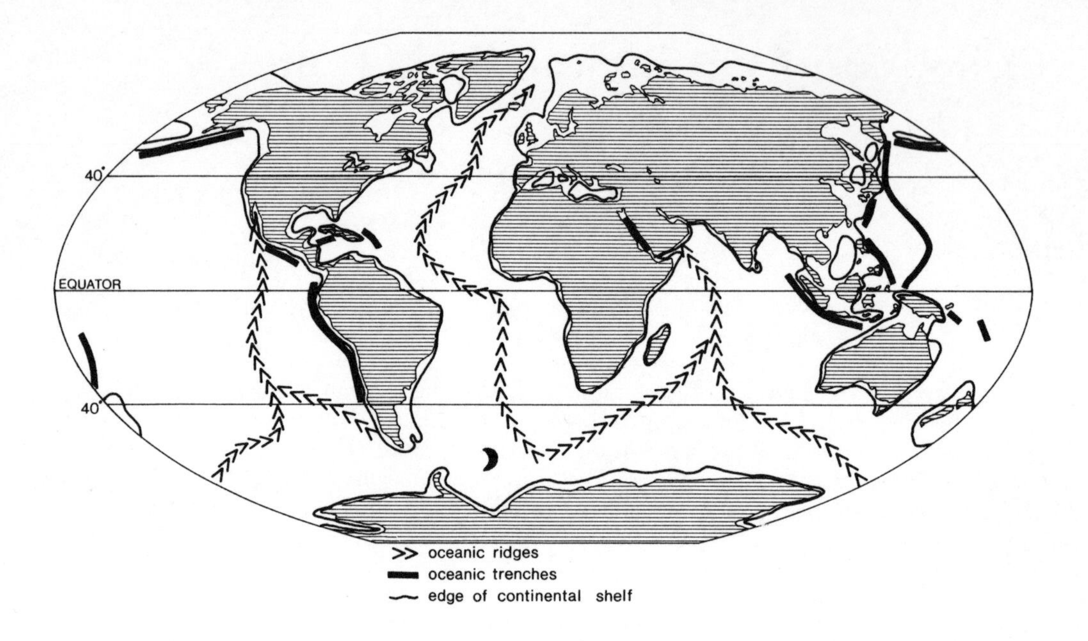

1.1 Major features of the ocean basins.

crust, which includes the continental shelves at present covered by up to 200 m of water (Fig. 1.1). The two are separated by the continental slope, which has a relatively steep incline, typically between 1 in 10 and 1 in 20. The coastlines which separate the land from water are very ephemeral features, changing considerably when water is either withdrawn from the ocean and locked up in continental ice sheets or returned to the ocean when these ice sheets melt. During glacial periods of the Pleistocene Epoch, such *eustatic* changes lowered sea level by some 100 m below its present level, and it is estimated that if all of the ice now on the Antarctic continent and Greenland were to melt, sea level would rise by about 60 m over the entire Earth. At the present time some 70% of the Earth's surface is covered by water, but only between 60 and 65% can be considered to be oceanic crust.

A notable feature of the ocean basins is that they are all interconnected. The Pacific, Atlantic and Indian Oceans radiate like spokes from the hub of the circumpolar Southern Ocean. The Arctic Ocean and European sub-Arctic seas may be considered as a large mediterranean sea, semi-enclosed by land and separated from the adjacent ocean basins by submarine ridges. In recent years the topographic features of the ocean basins have been shown to be related to their formation and development. This process, which has been called *sea-floor spreading*, is now incorporated in a concept known as *plate tectonics*. Oceanic crust originates from material in the Earth's mantle which finds its way to the surface in zones where crustal material is moving apart. It forms the ocean ridges which exist around the world ocean (Fig. 1.1), standing some 2 to 3 km above the surrounding ocean floor. In the southern Atlantic Ocean it has been shown that the sea floor has been spreading laterally at about 2 cm yr^{-1} on either side of the Mid-Atlantic Ridge for at least the last 80 million years, and thus the width of the ocean basin has been growing at the rate of about 1 km per 25 000 yr. Such active ridges are characterised by volcanic activity, a median rift valley, and a series of faults running at right angles to the crest and off-setting it by typically tens of kilometres. These features, which have been called *transform faults*, result from the relative movements of the quasi-rigid aseismic plates shown in Figure 1.2 which appear to comprise the uppermost 100–150 km of the Earth.

As well as providing the source of new crust on the Earth, the ocean basins provide the sinks at

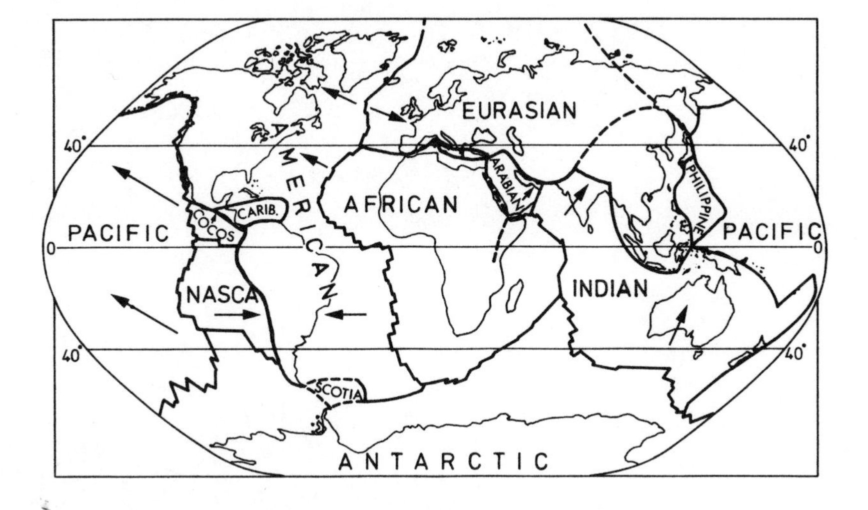

1.2 The extent of the major aseismic plates at the Earth's surface. (After M. G. Gross)

which crustal material is resorbed into the Earth's mantle. These are the oceanic trenches shown in Figure 1. 1 with greatest depths more than 10 km below sea level. Here two of the quasi-rigid plates meet, and one of them plunges down to form the trench. Such a situation occurs along the eastern side of the South Pacific, and here the leading edge of the westward-moving plate on which South America is located has been deformed to create the Andes mountains. Where two converging plates bring together oceanic crust, as in the western Pacific, volcanic activity associated with one of the plates descending has led to the formation of chains of volcanic islands such as the Philippines paralleling the oceanic trench.

The other main features which stand above the ocean floor (Fig. 1. 0) are abyssal hills and seamounts, almost all clearly of volcanic origin. Some of the seamounts known as *guyots* have flat tops, and even though they may now be 1 or 2 km below sea level they are assumed to have once been volcanoes reaching to the ocean surface, where they were truncated by wave action. Most of these features were presumably formed when they were within the active region of an ocean ridge, and have been carried to their present locations by the 'conveyor-belt' action of sea-floor spreading.

Much of the ocean floor comprises abyssal plains which are very flat, being covered by a thick layer of *sediment*. Some of this sediment is truly oceanic in origin, e.g. the various oozes which result from the decay of oceanic organisms. Clays found in the deep ocean are formed of very fine particles which have been transported by ocean currents, or winds, before settling out on to the ocean floor. Around the continental margins, much coarser sediments of terrestrial origin are found. These are contributed by rivers, glaciers, and by coastal erosion, and accumulate on continental shelves where they are re-worked by wave action and tidal currents into a variety of depositional patterns, e.g. off-shore bars and banks, sand waves and ribbons. When such an accumulation produces an unstable slope at the edge of the shelf, or perhaps when an earthquake occurs, a submarine avalanche known as a *turbidity current* develops down the continental slope. Such flows will be channelled by any existing declivities, and so tend to erode them further, leading to the formation of canyons in the continental slope. The sediment is kept in suspension by the turbulence in the flow while it is moving fast enough down the slope – speeds of 7.5 m s^{-1} and higher have been deduced for a turbidity current which broke a series of transatlantic cables on the continental slope off the Grand Banks of Newfoundland, and theoretical considerations confirm that such speeds are possible. But when such a current reaches an abyssal plain it spreads out and slows down, depositing its sediment load and thereby blanketing any irregularities on the ocean floor.

2. 0 A modern survey ship, H.M.S. Hecla, with a weather balloon tethered to her, making observations as part of the Global Atmospheric Research Project in the Atlantic.

2 Fluids in General: The Atmosphere and Water in Particular

Compressibility and hydrostatic pressure □ Composition of the atmosphere □ Adiabatic changes □ Viscosity □ The peculiar properties of water.

MATTER CAN BE IN ANY OF THREE STATES: solid, liquid or gaseous. The state in which a particular substance exists depends on the physical conditions prevailing at the time. Thus water usually occurs on the Earth's surface as a liquid, but in cold regions it takes the form of ice, and in the atmosphere it is present as water vapour (a gas) or as a fine suspension of droplets or ice crystals. Liquids and gases have the common property of deforming continuously (or changing their shape readily) under the action of shear stresses because the molecules of which they are composed are able to move easily relative to one another. They are therefore able to flow freely and are given the name *fluids*.

COMPRESSIBILITY

When fluids are subject to increased pressure (e.g. by the action of a piston on a sealed container) the molecules are forced closer together, the volume of the fluid decreases, and its density increases. We may distinguish fluids which are readily compressible from those which are virtually incompressible according to the amount by which their volume changes for a given change in pressure. In general, gases which are of relatively low density are readily compressible, whereas fluids which are considerably denser are virtually incompressible. If a gas is introduced into an empty, closed container it tends to expand to occupy it uniformly, in contrast to a liquid which in the same situation comes to rest with a free horizontal surface. These characteristics lead to the very different natures of the upper boundaries of the ocean and atmosphere – that of the ocean is quite clear and distinct, while that of the atmosphere is diffuse and impossible to define precisely.

In both the atmosphere and the ocean pressure increases proportionally with depth due to the increasing weight of the overlying fluid. The change in pressure Δp associated with a change in height (or depth) Δz is given by the hydrostatic equation:

$$\Delta p = -g\rho\Delta z \,, \qquad (2.\text{I})$$

where g is the acceleration due to gravity and ρ the density of the fluid, assuming that these both remain constant. (The negative sign indicates that we are taking z positive upwards.)

In the ocean this is a reasonable assumption – the density of all but about 1% of ocean water lies within about ±2% of its mean value, and the variation in g is very much less than this. Hence a graph of Δp against Δz in the ocean is approximately linear, and if we take g as 9.8 m s^{-2} and ρ_w as 1.03×10^3 kg m^{-3} we find that a column of sea water of height 10 m exerts a pressure of approximately 10^5 N m^{-2} or 1 000 mb (i.e. approximately one atmosphere).

In the atmosphere, on the other hand, the air is readily compressible, so that its density varies significantly with height. We must therefore consider the differential equation

$$dp/dz = -g\rho \,, \qquad (2.\text{II})$$

which states that the rate at which pressure changes with height falls off in proportion to the density at that height.

The behaviour of air in the atmosphere approximates closely to that of an 'ideal gas', i.e. one in which collisions between molecules are perfectly elastic and the molecules themselves occupy no space at all. We may therefore substitute for ρ in Equation 2. II using the Ideal Gas Equation:

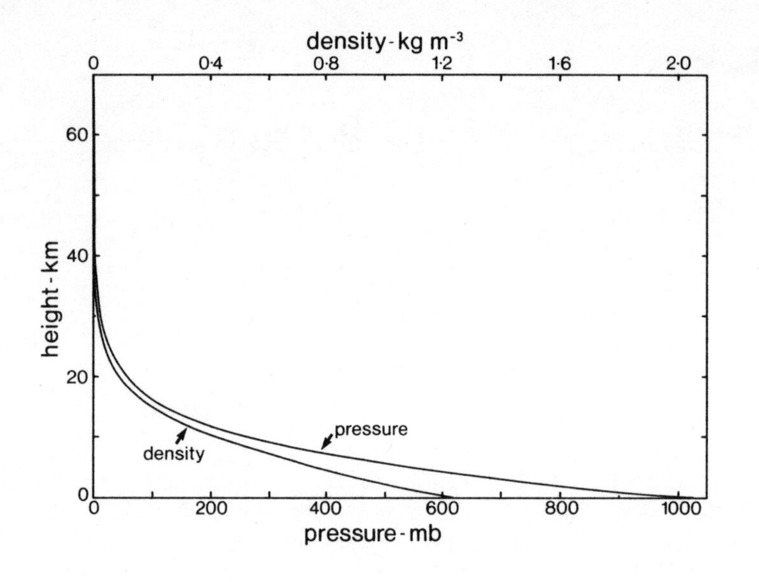

2. 1-a Typical variation of pressure and density with height in the atmosphere in middle latitudes.

$$pv = RT , \quad (2. \text{III})$$

i.e. pv/T is constant ,

where p, v and T are the pressure, volume and temperature of a given amount of the gas, and R is the universal gas constant. (This expression implies that as either p or v approaches zero while the other remains constant, T will approach zero, and this leads to the definition of an Absolute Zero of temperature which is used as the zero in the Kelvin scale of temperature measurement. At this temperature the molecules of a gas are theoretically stationary and therefore exert no pressure. The size of each degree in the Kelvin scale is the same as that in the Centigrade scale, but the zero is displaced to −273°C. Thus, for example, 283 K = 10°C. R has the value 8.3 J mol^{-1} K^{-1} .)

Putting M for the molecular weight of the gas, we have $\rho = M/v$, and from Equation 2. III

$$v = RT/p ,$$

so that

$$\rho = \frac{M}{v} = \frac{pM}{RT} . \quad (2. \text{IV})$$

Hence, writing R' for R/M,

$$\frac{dp}{dz} = -\frac{gp}{R'T} . \quad (2. \text{V})$$

Integrating Equation 2. V gives

$$\int \frac{1}{p} dp = \int -\frac{g}{R'T} dz ,$$

whence

$$p = p_0 e^{-gz/R'T} , \quad (2. \text{VI})$$

where p_0 is the pressure at $z = 0$.

Figure 2. 1 shows that the decrease of pressure with height in the atmosphere is essentially exponential, though the relationship is affected by variations of temperature with height (Chap. 6); as air density is directly proportional to pressure it also decreases approximately exponentially with height.

PRESENT COMPOSITION OF THE ATMOSPHERE

To some extent the composition of the atmosphere varies as a function of height. The lighter molecules migrate upwards, so that between 100 km and 1 000 km above the Earth's surface the atmosphere comprises mainly atomic oxygen; between 1 000 km and 2 400 km there is a layer of helium, and above 2 400 km hydrogen predominates. Solar radiation is responsible for other variations with height in the upper atmosphere

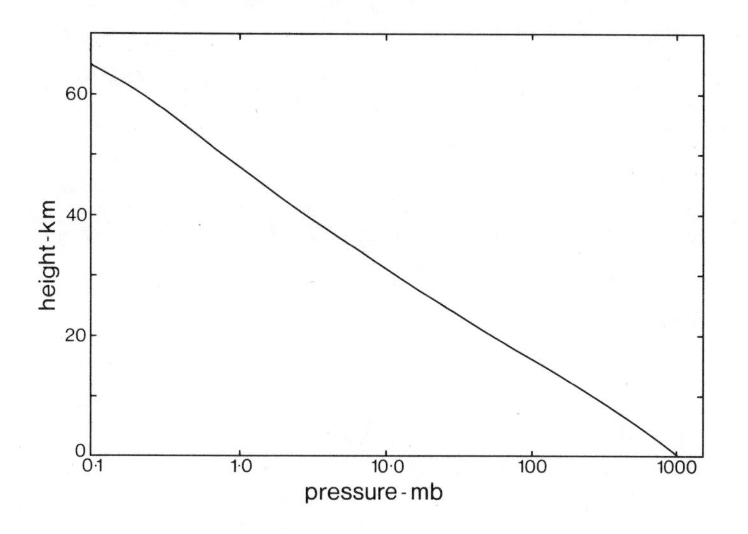

2. 1-b The variation of pressure with height plotted on a logarithmic scale.

(Chap. 5), but three-quarters of the mass of the atmosphere is concentrated in its lowest 10 km, and this part of the atmosphere shows no variations in the percentage concentrations of its major constituents: nitrogen (78%), oxygen (21%) and argon (1%). (All concentrations of atmospheric constituents are given by volume.) Apart from water vapour, to which we shall return later, there is one other constituent gas which is very important. This is carbon dioxide, which is fairly uniformly distributed in the lower atmosphere but accounts for only about 0.03% of the total at the present time. We shall consider its rôle in the oceans in Chapter 4 and in the Earth's heat budget in Chapter 5.

Fuel combustion and industrial processes lead to local concentrations of pollutant gases such as sulphur dioxide and ammonia as well as carbon dioxide, and of suspended particles. The latter also originate in dust storms, volcanic eruptions and salt spray over the oceans. Water is of course present in both its liquid and solid states as suspended drops and ice particles in clouds and fog.

ADIABATIC CHANGES

Another consequence of the compressibility of fluids is that *adiabatic* changes (i.e. changes which occur without any transfer of heat to or from the surroundings) take place. According to the First Law of Thermodynamics, a change under adiabatic conditions in the internal energy of a fluid (which determines its temperature) equals the external work which the fluid does in expanding or which is done on it when it is compressed. Thus as air rises it expands and loses internal energy, so that its temperature falls. This happens at the constant rate of 9.8°C km^{-1}. This *lapse rate* applies to air which is not saturated with water vapour, and is therefore known as the *dry* adiabatic lapse rate.

Adiabatic temperature changes are very much smaller in liquids which are relatively incompressible. In sea water the adiabatic lapse rate increases with both temperature and pressure, but is generally less than 0.2°C km^{-1} in oceanic conditions. In both the oceans and the atmosphere the temperature which the fluid would attain if brought adiabatically to a pressure of 1 000 mb (approximately sea level) is defined as the fluid's *potential temperature* (θ). The potential temperature of air may thus be some tens of degrees C higher than its *in situ* temperature, while the potential temperature of sub-surface water in the oceans is always lower than its *in situ* temperature, but never by more than 1.5°C.

VISCOSITY

A further property of fluids is *viscosity*, which is a measure of the tendency of two layers in a fluid to resist slipping over one another. It may be defined as the tangential force per unit area required to produce a unit velocity gradient in the fluid normal to the flow, and is measured in N s m^{-2}. A fluid with a high viscosity would be described as 'sticky', e.g. glycerine, and would not flow easily, whereas gases with much lower viscosities flow much more readily. Table 2. 1 gives the densities and viscosities of air, water and glycerine at atmospheric pressure and 20°C. (If there is irregular or turbulent motion, as in the atmosphere and ocean, the effective viscosity is many times larger than the values given below, which apply only to laminar flow.) The fact that air has a very much lower density and viscosity than water means that it is more easily set in motion, reaches higher velocities more rapidly, but comes to rest more quickly than water.

Table 2. 1

	DENSITY (kg m^{-3})	VISCOSITY (N s m^{-2})
air	1.2	1.8×10^{-5}
water	1.0×10^{3}	1.0×10^{-3}
glycerine	1.3×10^{3}	8.3×10^{-1}

WATER

Water, as well as being the main constituent of the oceans, is a very important constituent of the atmosphere, and it has been suggested that for meteorological purposes air can be regarded very largely as diluted water vapour. The concentration of water vapour in the atmosphere is highly variable, being greatest near the surface and in low latitudes, where it may account for more than 3% of an air sample over the tropical oceans. In its liquid form water is the commonest fluid in nature. Most organisms which have evolved on Earth are intimately dependent on it, and its properties have had a profound influence on almost all aspects of our environment. Modern life requires water in ever increasing quantities for domestic, industrial and agricultural uses. But despite its abundance and importance, water is a very peculiar substance.

The freezing and boiling points of a substance are related to the size of its molecules – the larger the molecules the higher these are. From such considerations, and comparison with other compounds of hydrogen, we would expect water to freeze at about –100°C and boil at about –80°C, in which case almost all water at present Earth temperatures would exist in the gaseous state. When a substance freezes, it is usual for its molecules to become more closely packed together, and so for its density to increase, but the density of ice is *less* than that of liquid water (see Table 2. 2). Thus freezing of water in cracks in rocks leads to shattering and to the breakdown of the rock into fragments; and the formation of ice at the top rather than the bottom of lakes and the ocean in high latitudes provides a layer of low thermal conductivity, insulating the water from further heat loss. Again, when a liquid is heated its molecules increase their kinetic energy and usually are at a greater mean distance from each other, thereby leading to a decrease in the liquid's density. In the case of *fresh* water, however, density increases as temperature rises from 0°C to 4°C, when its density is at a maximum value; from 4°C to boiling point its density decreases as temperature rises, as would be expected. One would

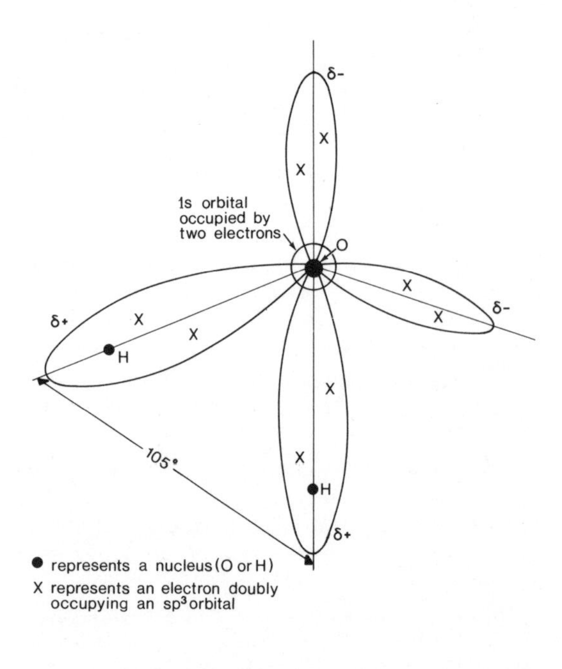

2. 2 Two-dimensional diagrammatic representation of the 3-dimensional structure of a water molecule. (After P. S. Liss)

predict that the viscosity of a liquid decreases as it is warmed, and increases as it is compressed – the closer the molecules are together the greater their resistance to flow – and this is so for most liquids. In water, viscosity decreases rapidly with rising temperature, and at low temperatures it decreases as pressure rises.

To find the reasons for these anomalous properties of water we must look at its molecular structure. The hydrogen atoms in a water molecule share electrons with the oxygen atom as shown in Figure 2. 2. The result of this sharing is that there is a net positive electrical charge in the vicinity of each hydrogen atom and a net negative charge associated with the oxygen atom. The four bonding arms of the oxygen atom make a 3-dimensional tetrahedral shape with angles between them of 120°. In the water molecule the repulsion between the negative charges of the unshared electrons pushes together the bonding arms on which the electrons are shared with the hydrogen atoms, reducing the angle between these arms to about 105°. Thus one side of the water molecule on which the hydrogen atoms are located has a small excess positive charge, while the other side has a small excess negative charge. This gives water molecules an attraction for one another so that they arrange themselves into partly ordered groups or structures, a process known as *polymerisation*. These bonds between water molecules are called hydrogen bonds.

As the temperature of water is raised, the energy of the molecules increases and they are able to break these hydrogen bonds and separate from the groups. When they do so they can fit between the groups and thus occupy less space, leading to a higher density. It seems that a balance between this effect and the normal expansion of a substance with increasing temperature is achieved at 4°C in fresh water – below 4°C this effect dominates, but above 4°C the normal thermal expansion exceeds it. Similarly the decrease in viscosity with increasing pressure at low temperatures suggests that water has a structure at these temperatures which impedes flow, but which can be crushed out by increased pressure.

Table 2. 2
Density of fresh water and ice at various temperatures

TEMP. (°C)	STATE	DENSITY (kg m^{-3})
−2	Solid	917.2
0	Solid	917.0
0	Liquid	999.8
4	Liquid	1000.0
10	Liquid	999.7
25	Liquid	997.1

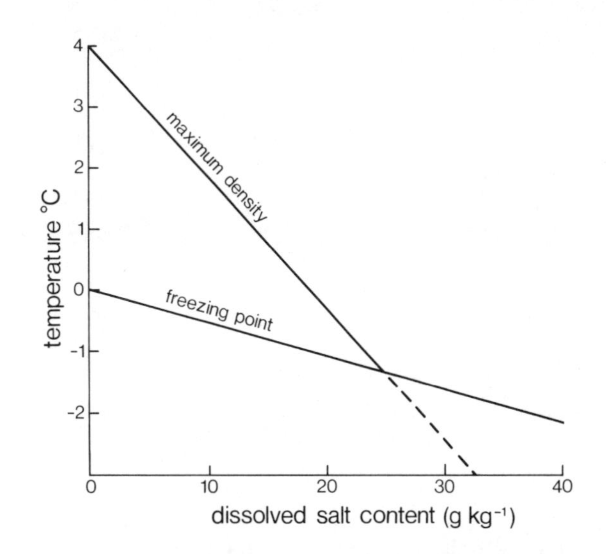

2. 3 Temperatures of freezing point and maximum density of liquid water as functions of dissolved salt content.

Water is an extremely effective solvent, a property which can be explained at least in part by the separation of electrical charges on water molecules. The individual ions in solids (e.g. the sodium cation Na^+ and the chloride anion Cl^- in sodium chloride) attract the opposite electrical charges on the water molecules, and the individual ions are surrounded by water molecules in a hydrated layer. This, however, disrupts the water structure, and modifies some of its physical properties. Its freezing point is depressed (an effect which is utilised when roads are sprayed with salt to melt ice) and the temperature at which it reaches its maximum density is reduced (Fig. 2. 3). When the salt content of the water reaches some 25 gm kg^{-1} the temperature of maximum density and the freezing point coincide at about −1.3°C, and if the salt content is higher than this (as in the ocean), the density of water increases as temperature decreases right down to freezing point.

Other exceptional and important properties of water are its high surface tension, specific heat and latent heats of fusion and vaporisation. Surface tension is a measure of the force needed to rupture the surface of a liquid. It is due to the cohesive forces between the molecules of a liquid, and so not surprisingly it is particularly high for water. The only substance which is a liquid at normal Earth surface temperatures which has a higher surface tension than water is mercury. The surface tension of water is important in the formation of drops in the atmosphere and of very small capillary waves on the ocean surface, and in the capillary transport of water in soils and organisms.

Its high specific and latent heats are vitally important in storing thermal energy and preventing extreme ranges in temperature. Its specific heat, which is the quantity of heat required to raise the temperature of a unit mass of it through one degree centigrade, is the highest of all solids and liquids except ammonia. Much of this heat energy is held in the bonds between adjacent water molecules which may be considered to act like springs holding the molecules together. As more energy is added to water, the molecules vibrate more and the temperature rises, but this is greatly restricted by the action of the 'springs'. Eventually the molecules may break loose completely from these bonds and thus migrate from liquid water to its gaseous state, water vapour. To break the 'spring' completely, however, requires an enormous amount of energy – at 20°C the same amount of energy could either increase the temperature of 585 kg of water by 1°C or evaporate 1 kg of water. This latent heat of evaporation is the highest of any substance. If the water vapour subsequently returns to the liquid state its latent heat is released. For water to freeze and form ice, heat must be extracted from it, so that the molecules possess considerably less energy and take their place in the regular ice lattice structure. As bonding exists between adjacent water molecules in both the ice and the liquid state, less latent heat is involved in this transformation than in that from liquid to water vapour. The amount of heat which is required to melt 1 kg of ice could increase the temperature of only 80 kg of water by 1°C – still more than for any substance except ammonia.

3 The Hydrological Cycle: Water in the Atmosphere

Stages and processes of the hydrological cycle □ Evaporation □ Saturation □ Vapour pressure □ Dewpoint □ Relative humidity □ Condensation □ Fog □ Cloud formation □ Cloud types □ Growth of cloud droplets □ Precipitation.

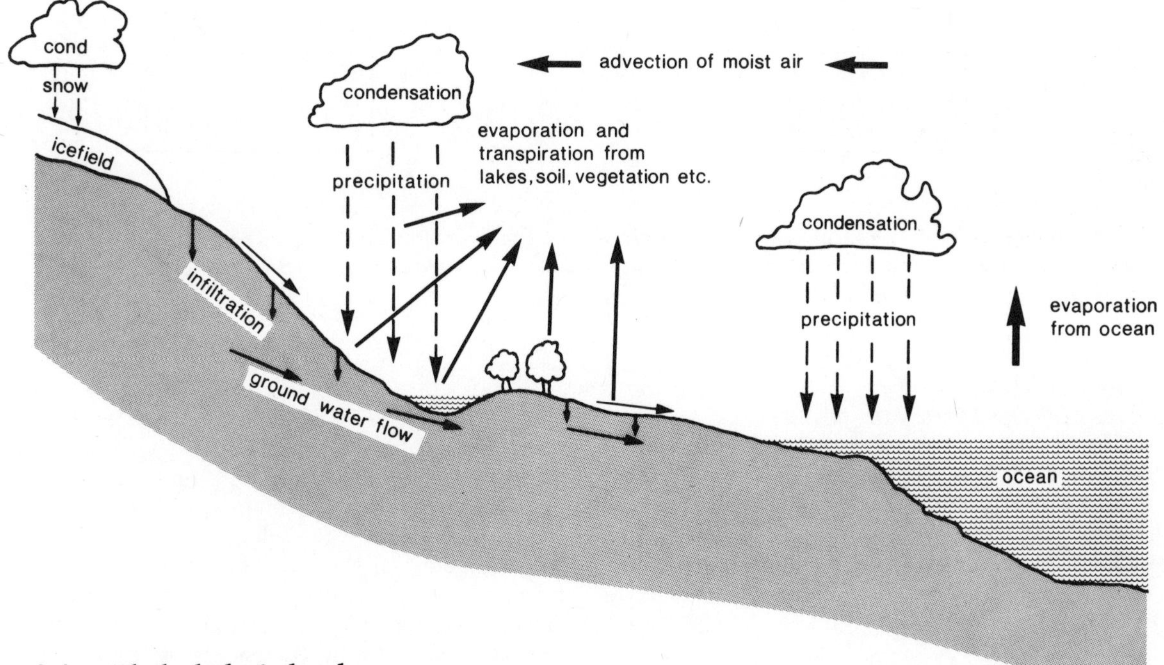

3. 1-a The hydrological cycle.

THE MAIN FEATURES of the hydrological cycle are well known and are summarized in Figure 3. 1-a. In it, water is continually changing from one state to another, and from one part of our environment to another. We may consider the cycle as comprising a series of stages and processes (Fig. 3. 1-b). The amount of water entering any one stage must equal the amount leaving it if a steady state is assumed, and the average length of time which a water molecule spends in any one stage (i.e. its *residence time* there) can be estimated by dividing the mass of water in that stage by the rate at which it is removed from it, or added to it, by whatever processes may be involved. The results of such estimates, taking the quantities in Figure 3. 1-b and the total mass of water involved as $1\,400 \times 10^{18}$ kg, give the following orders of magnitude for residence times: in the oceans 4×10^3 yr; in groundwater, ice, lakes and rivers combined 4×10^2 yr; in the atmosphere (water

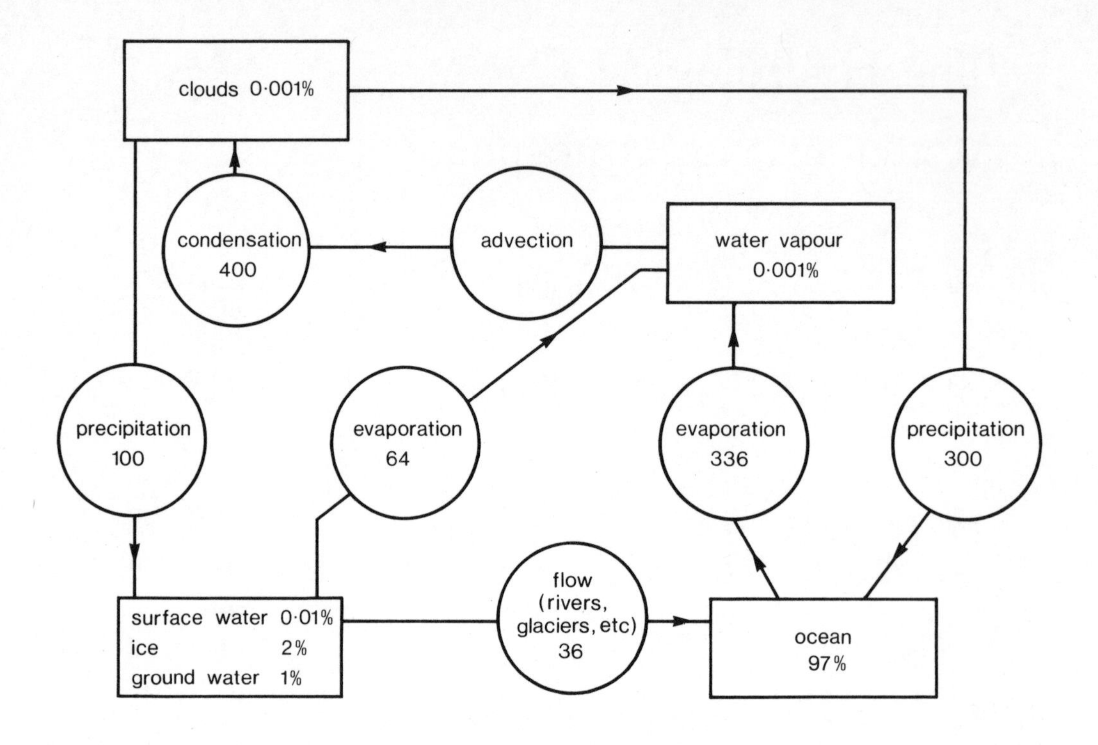

3. 1-b Stages (in rectangles) and processes (in circles) of the hydrological cycle. The proportion of the total water in the cycle which is present in each stage is shown as a percentage, and the rates of the processes are given in units of 10^{15} kg yr^{-1}.

vapour and clouds) 10 days. In fact water is being added very slowly from the *juvenile* water within the Earth which reaches the surface at volcanoes, thermal springs etc., and it is often very difficult to distinguish groundwater participating in the hydrological cycle from this juvenile water, making the estimate for the amount of groundwater in Figure 3. 1-b very uncertain.

EVAPORATION

The process whereby water is transferred from the ocean and from land surfaces into the atmosphere is known as evaporation. When it occurs from plant surfaces it is called transpiration, and when it occurs directly from an ice surface to the vapour state it is sublimation. The water vapour which is thus added to the gases in the atmosphere increases the pressure within the atmosphere.

Consider a closed container which is initially half filled with water overlain by dry air containing no water vapour. The molecules in the water will be in random motion, the kinetic energy of which is governed by the temperature of the water. They will collide with one another, thus transferring energy, and some near the liquid surface will achieve sufficiently high upward velocities to escape from the restraining attractive force of the other water molecules, and thus become water vapour. If the level of water in the container is kept constant by an external reservoir arrangement, the volume occupied by the air will remain constant whilst its mass is increased by the addition of water vapour. This leads to an increase in the pressure exerted by the air on the walls of its container, and the part of the total pressure which is attributable to the water vapour is referred to as the vapour pressure (e). This is a very convenient way of specifying the amount of water vapour present in a given sample of air. An alternative parameter is the *humidity mixing ratio*, which is the ratio of the mass of water vapour to the mass of dry air.

The molecules of water vapour will move around rapidly in the air above the liquid, and some will strike the water surface and be retained by it, returning to the liquid state – the process of condensation. If the system is left for a sufficient length of time, equilibrium will be achieved when the processes of condensation and evaporation just balance one another, and from then on the amount of water vapour in the air remains constant. The air is then said to be saturated with water vapour, and the pressure which this water vapour exerts is said to be the *saturation vapour pressure* with respect to a water surface (e_w). Not surprisingly, since the temperature controls the kinetic energy of the water molecules, the saturation vapour pressure is very temperature-dependent, increasing more and more rapidly as temperature increases (Fig. 3. 2). Note that below 0°C the saturation vapour pressure is less over an ice surface than over a supercooled water surface. (Liquid water can be cooled well below 0°C without freezing unless particles which serve as ice nuclei are present for the ice crystals to form on.) When there is no surface of any kind for water vapour to condense on, air can become highly supersaturated and still retain its water vapour.

Using Figure 3. 2, one can obtain two further measures of the amount of water vapour present in a sample of air, which give an indication of the temperature at which condensation will take place and of the additional amount of water vapour which the air could hold. The first is the *dew-point* temperature, defined as the temperature at which the air sample would reach saturation with respect to a water surface if it were cooled at constant pressure. The equivalent temperature with respect to an ice surface is the *frost-point*. In the example in Figure 3. 2, to find the dew-point of sample A (temperature 25°C, vapour pressure 20 mb) we follow the line AB and read off the temperature of point B (17.5°C).

The second of these measures is the *relative humidity*, which is given by

$$U = 100\% \times e/e_w \,. \qquad (3.1)$$

In the example above, e is 20 mb, and e_w (the saturation vapour pressure at the air temperature of 25°C) is 31.5 mb (point C), so that $U = 100\% \times 20/31.5 \simeq 63.5\%$. The relative humidity is increased not only by an increase in the water vapour content, but also by a decrease in temperature if the water vapour content remains constant, and so the diurnal variation in relative humidity often mirrors the diurnal variation in air temperature.

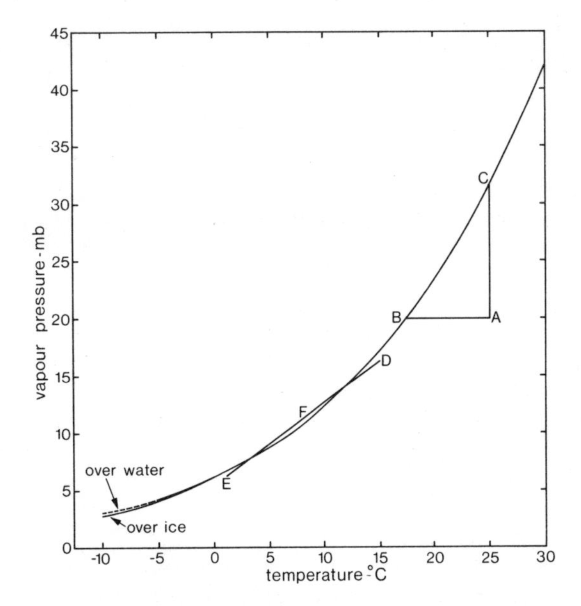

3. 2 Saturation vapour pressure as a function of temperature.

CONDENSATION

Although neither water nor ice surfaces are available in a free cloudless atmosphere, there are many impurities such as salt particles from the evaporation of sea spray, dust from deserts and volcanic eruptions and smoke from fires on which condensation can take place. These are known as *condensation nuclei*. They vary in their effectiveness in promoting condensation, but there are usually ample nuclei in the atmosphere for condensation to occur when the relative humidity just exceeds 100%. Nuclei on which condensation occurs below a relative humidity of 100% are known as *hygroscopic nuclei*, and are usually soluble salt particles or industrial pollutants. Condensation can also form on the ground as dew or, when the temperature is less than 0°C and sublimation occurs, as hoar frost.

The saturation of air leading to condensation usually results from the air being cooled. There is one other important process, which is illustrated in Figure 3. 2. Consider the samples of air represented by points D and E. Neither is saturated with water vapour, but if they are thoroughly mixed together in equal quantities the resultant mixture will be represented by point F – which is saturated. Hence it is seen that mixing of two different types of air can lead to saturation and condensation.

The more usual way by which air becomes saturated, however, is by cooling, which may result from contact with a cool surface or ascent of the air. If the air is in contact with a surface colder than its dew-point and is almost still (wind less than a few knots), dew or frost will form; but if there is a light wind, the cooled air is mixed in a shallow surface layer and, if the cooling is sufficient, all of this layer may become saturated with water vapour and fog develops. A stronger wind would mix a deeper layer of air, and this would be less likely to be all cooled down to its dew-point, and thus a strong enough wind prevents fog forming. By international convention, fog is said to occur when horizontal visibility is reduced to less than 1 km by water droplets. If visibility is similarly reduced by dust particles, the condition is described as haze, and if it is reduced by water droplets but not to less than 1 km it is described as mist. Smoke particles and industrial pollutants reduce the visibility both directly and indirectly by providing hygroscopic nuclei which promote condensation. Such a thick, mixed fog is known as *smog*.

Two main types of fog are distinguished according to the reason for the underlying surface being colder than the air:

1. *Radiation fog* results from the ground surface cooling by long-wave radiation into space at night. For reasons which will be discussed in Chapter 5 the sea surface cools down very little at night by comparison with a land surface, and cooling of a land surface by long-wave radiation is only effective when there are few clouds above it and the sky is relatively clear. Thus radiation fog forms over land surfaces on clear nights with a light breeze, and it is most likely when the lowest layers of air start with a high relative humidity and the surface is cold and wet, as for example over marshy areas in the winter half of the year. As the air in which the fog is present is cool and therefore relatively dense it will tend to flow downhill into hollows, and may drift seawards, particularly over estuaries. In most cases the Sun will penetrate such radiation fog fairly early in the morning, and will dispel it by warming; but if a thick layer has formed, or if the fog has drifted over a cool sea surface and it is winter, when the solar heating is weak, then it may persist all day.

2. *Advection fogs* result from the horizontal motion of air over a colder surface, either land or sea. They are most persistent over the sea, whose surface is less subject to rapid warming by the overlying air or by solar heating, and most sea fogs are of this type. When the temperature difference between the air and an underlying water surface is large, the fog is very persistent and extends to heights of perhaps 200 m above sea level when the wind is strong. This sometimes happens, for example, on the Grand Banks off Newfoundland, where air which may have come from over the warm Gulf Stream and so has a high relative humidity as well as a high temperature, is advected over the cold waters of the Labrador Current (Fig. 3. 3). Such fogs must be expected over any sea area where the surface temperature is low, whether this be due to a cold surface current or to the upwelling of cool water from below, and are most frequent in spring and summer, when the air is likely to be at its warmest relative to the sea and rich in water vapour. Over land, advection fogs are commonest in winter when warm, moist maritime air spreads over a cold land surface; but as a land surface can warm up fairly rapidly, the fog is soon dispelled unless the processes responsible for radiation fog intervene to retain it.

In addition to these two major types of fog a number of less important ones can be recognised. *Steam fog*, or 'arctic sea smoke' as it is sometimes called, occurs when very cold air travels across a sea surface and continuous rapid evaporation takes place. The air in contact with the sea gains heat as well as becoming saturated with water vapour with respect to the surface water temperature. It then rises and mixes with the colder air above which has a very low capacity for water vapour, so that a super-saturated mixture develops and condensation takes place. Such fogs are usually very shallow and, as they require minimum temperature differences between the water and the air of the order of 10°C, are most common at the edge of the pack-ice when the wind is blowing off the ice. *Frontal fog*, or mixing fog, is formed at the boundary between two different types of air, each near to saturation, by the process illustrated by the line *DEF* in Figure 3. 2. *Upslope fog* occurs on the side of a hill or mountain, and is really low cloud in contact with high ground, so that its formation will be better discussed under the heading of clouds.

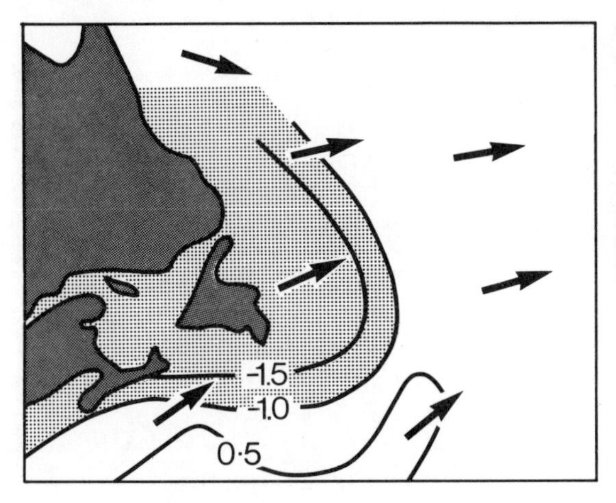

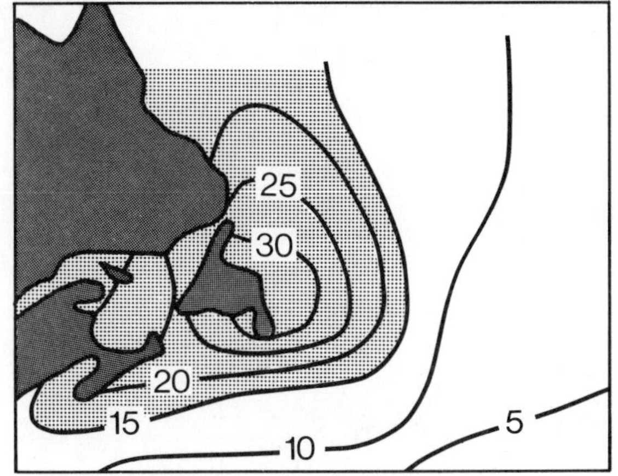

3. 3-a Sea-surface minus air temperature (°C), and the prevailing wind direction, over the Grand Banks of Newfoundland in March, April and May. (After H. U. Sverdrup et al.)

3. 3-b Percentage frequency of fog in the same months. (After H. U. Sverdrup et al.)

CLOUD FORMATION

Clouds are the result of condensation in the free atmosphere. Apart from the mixing process leading to saturation which has already been mentioned, condensation in the atmosphere results from air being cooled adiabatically to its dew-point by ascent (Chap. 2). There are three principal causes of air rising in the atmosphere:

When air which is moving horizontally encounters a hill or mountain barrier it must pass over or around it if it is not to pile up behind it. Unless the barrier is of very small lateral dimensions, the air will initially slow down and begin to pile up behind it, but the following air will then have to rise even higher to pass over this (Fig. 3. 4). If air has to rise 500 m over such an *orographic* barrier, and it does this without becoming saturated, it will cool down by 5°C. Probably the air at some levels will reach its dew-point with this amount of cooling, and clouds will form, but air at other levels will not. As shown in Figure 3. 4, oscillations known as *lee waves* may develop in the air flow downstream of the obstruction, and stationary lenticular (lens-shaped) clouds may form in them with water vapour condensing as

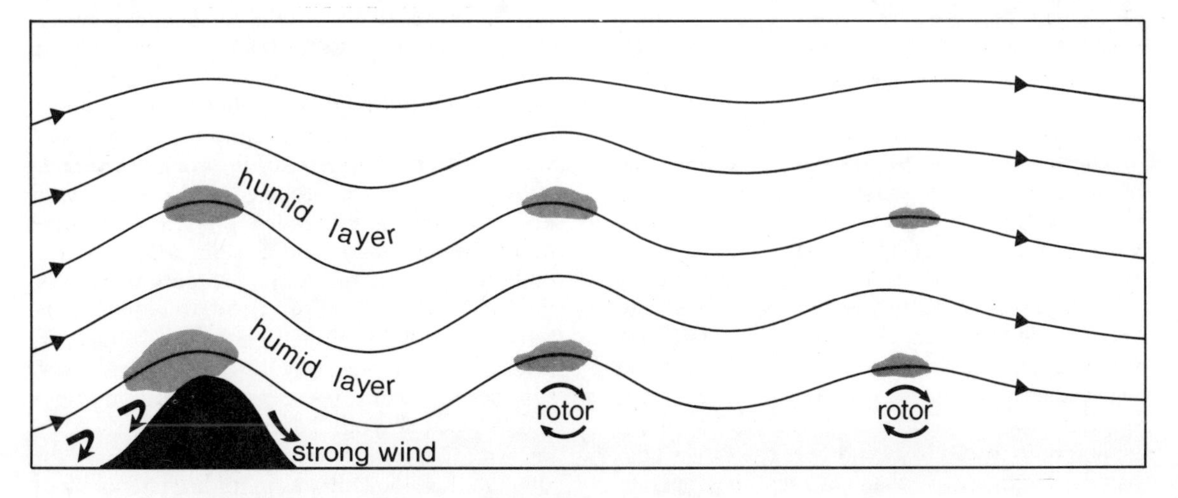

3. 4 The flow of air over an orographic barrier, and associated clouds and lee waves.

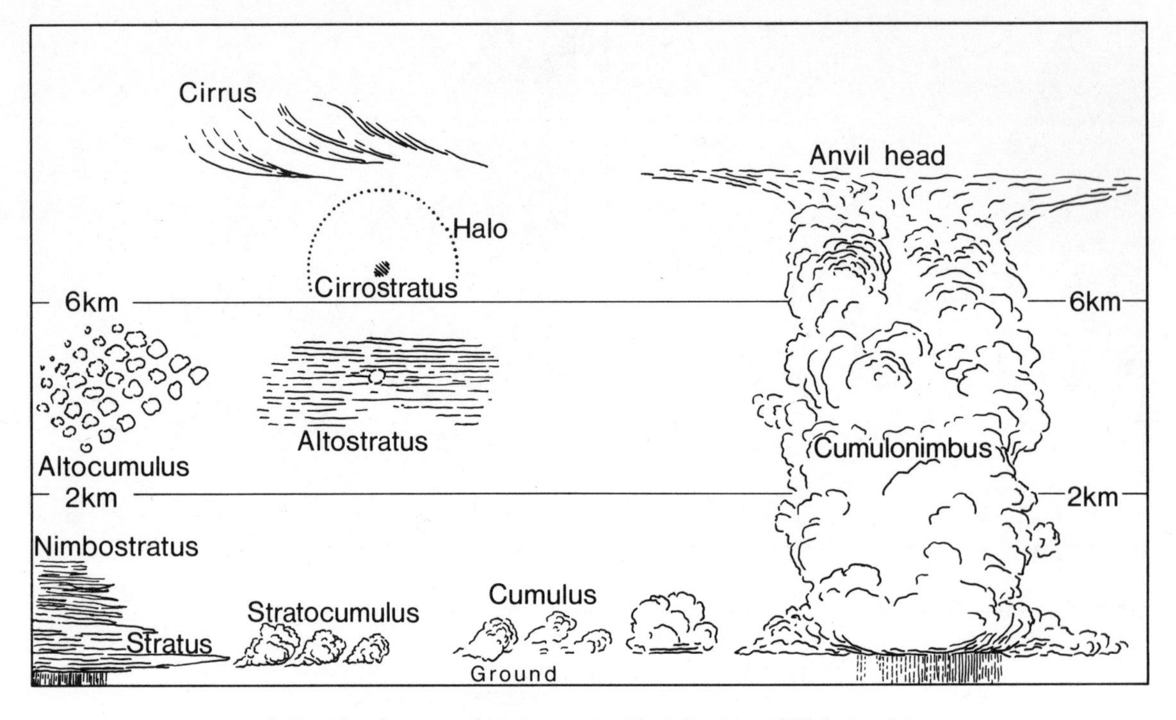

3. 5 Cloud genera showing typical heights in middle latitudes.

the air moves up in the wave and then evaporating as it moves down on the other side.

The second principal cause is the horizontal convergence of air. When this happens on a sufficiently large scale it is complicated by the Earth's rotation which, away from the equator, leads to the development of cyclones (Chap. 13). Often in middle and high latitudes it involves the meeting of two different masses of air, in which case the one which is less dense is forced to rise over the other. Such ascent of air along the boundary, or *front*, which separates different air masses is generally referred to as *frontal uplift*.

The third cause is the warming of air near the ground making it less dense and thus able to rise – the process of convection. We shall consider in Chapter 6 the factors which control the point at which the air column becomes unstable and convection starts, and the processes which control the height to which convection reaches. On a hot sunny day, air over a mountain might well be warmed sufficiently for convection to develop, but this process could be triggered off by the air having been forced to ascend over the orographic barrier, and thus more than one cause may operate in a particular situation.

CLOUD TYPES

The clouds which develop are classified according to their general shape, whether they comprise water droplets or ice crystals, the height of their bases and their vertical extent. Figure 3. 5 shows the main cloud genera which are recognised. The cirrus clouds are essentially ice clouds, and as ice crystals are relatively slow in forming and evaporating the shape of cirrus clouds gives an indication of the vertical shear of the wind at the height at which they are observed. Hooked cirrus, which has a characteristic shape, indicates a very strong vertical shear. Cirrostratus is distinguished from altostratus, a partly droplet-laden cloud, by its fibrous outline as distinct from the clear-cut edges of the water cloud, and by the halo which is observed around the Sun or Moon when seen through a thin layer of cirrostratus due to the refraction of light by the ice crystals. Stratus are layer clouds at relatively low levels, with their base below 2 km. If the stratus clouds are thick and associated with continuous rain, they are known as nimbostratus. Stratocumulus and altocumulus clouds are distinguished essentially by the height of their base, though stratocumulus is

3.6 Five common types of cloud formation.

Top left: the anvil plume resulting from wind shear acting on towering cumulonimbus.

Top right: low cloud extending over the English coastline at Ramsgate.

Middle left: altocumulus floccus (in centre) and altocumulus castellanus (bottom right).

Middle right: cirrus cloud formation associated with a jet stream.

Bottom: altocumulus in bands.

3.7 Multi-layered wave clouds formed in air flow over mountains of Graham Land. Taken from ca. 65°S, 65°W, looking north-east.

usually thicker and has less clear space between individual clouds than altocumulus. They may form from stratus and altostratus respectively as a result of turbulent or convective mixing. Cumulus may be, and cumulonimbus always is, of considerable vertical extent. They are both water droplet clouds at low levels, but the upper part of cumulonimbus clouds, which are associated with heavy showers, is glaciated, i.e. is made up of ice crystals. The *International Cloud Atlas* also includes cirrocumulus as a cloud genus; but this implies a water droplet cloud, whereas cirriform clouds are by nature ice clouds. Some authors therefore prefer to omit this genus and classify such clouds as cirrus, cirrostratus or altocumulus, as is most appropriate. Further details of cloud species and varieties may be found in the *International Cloud Atlas*.

The different cloud genera can be associated with the ways in which they are formed. Steady ascent over a wide area due to orographic effects or frontal uplift leads to the formation of the stratus-type cloud appropriate to the altitude. Convection leads to the formation of cumulus and cumulonimbus clouds, and turbulent mixing may lead to the formation of stratus clouds or (more often) stratocumulus clouds if the turbulence is in the surface layer, and to altocumulus clouds if it is in a higher layer, perhaps induced there by a strong vertical shear in the wind.

The droplets which make up a typical water cloud are very small, having an average radius of about 10 μm. Such drops have a fall speed through still air of about 0.01 m s^{-1}, and thus remain suspended at about the same level if the air is slowly ascending at this sort of speed. If the air ceased to ascend, it would take a cloud droplet of twice the average radius roughly half an hour to reach the ground from a cloud base just 100 m above the surface, and this would give more than ample time for all of the drops to evaporate before reaching the ground.

The raindrops which reach the surface are typically about 1 mm in radius, and thus contain 10^6 times as much water as the average cloud droplets. For a cloud droplet to grow to this size only by the process of condensation would take some hours, and there would usually be insufficient water vapour available in the air to allow more than a small proportion of the droplets in a cloud to grow to this size. As precipitation often develops within an hour in a cloud, there must be processes at work which allow some drops to grow at the expense of others, or which lead to drops coalescing. Much research has been devoted to identifying these processes, and many hypotheses have been tested. There are two theories which appear to account for cloud droplet growth on most occasions.

The first theory uses collision between droplets to explain their growth. Random turbulent motion within a cloud will cause droplets to collide occasionally, but if they are both of the same size they are just as likely to break up as to coalesce. However, if a cloud contains droplets of varying sizes (due perhaps to differences in the nuclei on which they originated rather than to collisions) the droplets will move relative to one another due to their different fall speeds. The larger droplets will have a downward motion relative to the smaller ones, and the process by which they encounter and absorb the smaller ones has been described as 'sweeping'. Such processes are most effective in tropical cumuliform clouds of considerable vertical extent with strong up-currents and high moisture content of the air.

The second, known as the Bergeron-Findeisen Theory, is based on the different values of saturation vapour pressure with respect to an ice surface and a supercooled water surface (Fig. 3. 2). Ice nuclei on which freezing will take place above $-30°C$ are not present in large numbers in the atmosphere, and in clouds at between $-12°C$ and $-30°C$ there is usually a mixture of ice crystals and supercooled water droplets. Because the saturation vapour pressure with respect to ice is less than that with respect to supercooled water at the same temperature, the air can be saturated with respect to the ice although the relative humidity is less than 100% over the water, and evaporation takes place. The ice crystals thus grow at the expense of the water droplets and soon become large enough to fall through the cloud. Their movement relative to the water droplets leads to collisions, and coalescence becomes effective. As they fall to the ground they may well melt in warmer air and so reach the surface as raindrops. Some, however, will fall as snow or sleet, and others will be caught in up-currents after melting, be frozen once again, and perhaps after this has happened a number of times will reach the surface as hail. This process is likely to be very effective in middle and high latitudes, but cannot account for precipitation which falls in the tropics from clouds which are at temperatures above freezing point throughout.

4.0 Top: at the edge of the Antarctic continent.
Left: tabular antarctic iceberg between South Orkney and Elephant Islands.
Right: old arctic iceberg, layered and tilted, off King Karl Land, Spitzbergen.

4 Salts, Gases and Ice in the Oceans

Constituents of sea water □ Salinity □ Nutrient salts □ Dissolved gases: oxygen and carbon dioxide □ Formation of sea-ice □ Icebergs.

MOST OF THE WATER taking part in the hydrological cycle is, at any one time, in the oceans, and has salts and gases dissolved in it. Virtually every element is probably dissolved in sea water, though not all have yet been identified in it and many are present in such small concentrations that very sophisticated analytical techniques are required to detect them. Some of these dissolved constituents are extracted commercially – apart from common salt (NaCl), magnesium and bromine are extracted on quite a large scale, and there have been many ideas for obtaining precious metals such as gold from sea water, but so far they have proved uneconomic due to the very low concentrations involved. The suspended particles, which gradually settle out of sea water and form oceanic sediments, are separated from the dissolved constituents by passing the water through a very fine filter (0.45 μm membrane filters are commonly used); the suspended particles are trapped while the dissolved constituents pass through with the water. These suspended particles reduce the transparency of the water to light (i.e. they increase its turbidity), and they raise its density, thus permitting turbidity currents to develop.

SALINITY

The easiest way to separate salts from the water in which they are dissolved would seem to be by evaporation. If we could do this, driving off all the water without any loss at all of the salts, we could readily find the weight of salt dissolved in a unit mass of sea water. (In fact, if sea water is heated to evaporate all its water content, some of the ionic constituents react to form gases which are lost to the atmosphere, and hence this is not a satisfactory method of determining the salt content.) We would typically find about 35 g of salt in 1 kg of sea water, with eight ionic constituents making up almost 99.9% of the total mass of salt (Fig. 4. 1). These can be considered as the major constituents. In the last century it was found that, although the total concentration of salt varies from place to place, the relative proportions of these major constituents are remarkably constant. The work of Dittmar (who in 1884 published the analyses of 77 samples collected on the world voyage of *H.M.S. Challenger*) confirmed this, and the most recent work to be carried out shows it to be true for all but appreciably diluted sea water for most practical purposes, though the ratio of the calcium content to the total salt content has been found to vary by up to about 0.5% of its mean value.

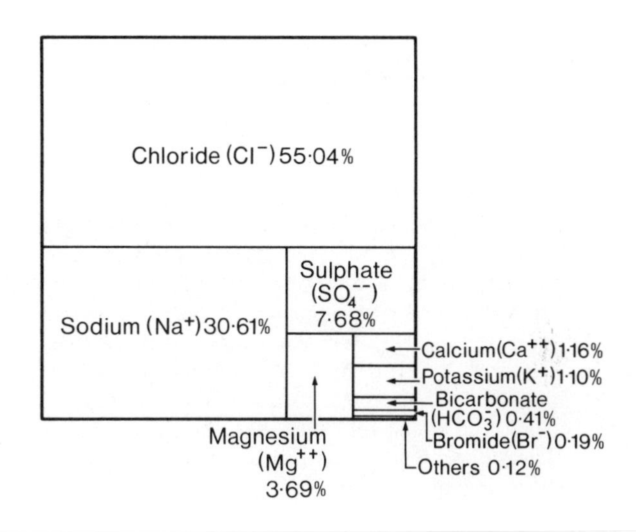

4. 1 The major ionic constituents dissolved in sea water.

Until about 1960 this law of constancy of composition provided the basis for the principal method of determining the total salt content of sea water, i.e. its *salinity*, which is expressed in gm kg^{-1} as ‰. This method determined the *chlorinity* of the sea water by titration with silver nitrate. The chlorinity is the mass of chlorine, bromine and iodine in a given mass of sea water which is precipitated out by silver, assuming that the small amounts of bromine and iodine involved have been replaced by chlorine. The main reaction is thus:

$$Cl^- + AgNO_3 \rightleftharpoons AgCl\downarrow + NO_3^- .$$

The silver chloride precipitate is white, and in order to determine quantitatively the amount of chloride ions present it is necessary to know the precise amount of silver nitrate which has to be added to just precipitate all of the chloride ions. The standard way of doing this is to add potassium chromate solution which, at the point when all of the chloride ions have been precipitated, reacts with the silver nitrate to form silver chromate, which has a distinctive red colour, and thus marks the end-point of the titration. To standardise the amount of silver present in the silver nitrate solution which is used, the titration is first carried out with a 'Normal (*or* Standard) Sea Water' prepared by an international agency in Copenhagen for which the chlorinity is already known. The salinity is then computed from the chlorinity using an empirical ratio:

$$S(‰) = 1.806\,55\ Cl(‰) .$$

In recent years the determination of salinity by chlorinity titration has been largely replaced by the measurement of electrical conductivity, using instruments known as *salinometers*. There are a number of properties of sea water which depend upon its salinity, and many of these have been used to determine it, e.g. refractive index for light, or velocity of sound in it. Electrical conductivity measurements are particularly suitable due to the speed with which they can be made and to the relative ease of transmitting electrical signals to a surface observer if *in situ* measurements are required. The main problems arise from the strong dependence of electrical conductivity on temperature and to a lesser extent on pressure, and from polarization at the electrodes where contact is made between the water and the instrument. The latter has been overcome by the use of induction techniques. The former at first required the use of large thermostatic baths, but can now be dealt with by compensation devices incorporated in the measuring circuit. Instruments for *in situ* measurements must either have such devices, or include provision for the separate measurement of temperature and pressure for subsequent numerical calculation of salinity from conductivity. Once again these instruments are standardised by the use of 'Normal Sea Water' from Copenhagen, and salinity has now been re-defined in terms of the ratio of the conductivity of a water sample to that of water having a salinity of exactly 35‰, both samples being at a standard temperature and pressure.

The salinity of the surface water of the oceans is maximal in latitudes of about 20°, decreasing towards higher latitudes and towards the equator (Fig. 4. 2). This is in response to the balance between the removal of fresh water by evaporation and its input by precipitation. The regions of highest salinity are those where evaporation exceeds precipitation, and correspond to the hot, barren deserts which exist in similar latitudes on land. In high latitudes the formation and melting of ice complicate the surface salinity distribution, and also bring about seasonal variations in it. In the deep waters of the oceans salinity shows relatively little variation – below a depth of about 2 km it is almost invariably between 34.5 and 35.0‰.

NUTRIENT SALTS

Of the minor constituents of sea water which do not appear in Figure 4. 1, some, namely the nutrient salts, are vitally important for the organic life of the oceans. These include phosphates, nitrates and silicates which are utilised by the phytoplankton, the minute plant organisms which drift in the sea and which can photosynthesize carbohydrates from carbon dioxide and water, e.g.

$$6CO_2 + 6H_2O \xrightarrow{\text{sunlight}} C_6H_{12}O_6 + 6O_2 . \quad (4.1)$$

To do this they must remain in the photic zone where there is sufficient sunlight, and even in relatively clear ocean water this is confined to the uppermost 100 to 200 m. It is in this layer, therefore, that nutrient salts are utilised. The phytoplankton forms the base of the oceanic food chain, and thus these nutrient salts move along this food chain as predation takes place, generally moving downwards below the photic zone as they pro-

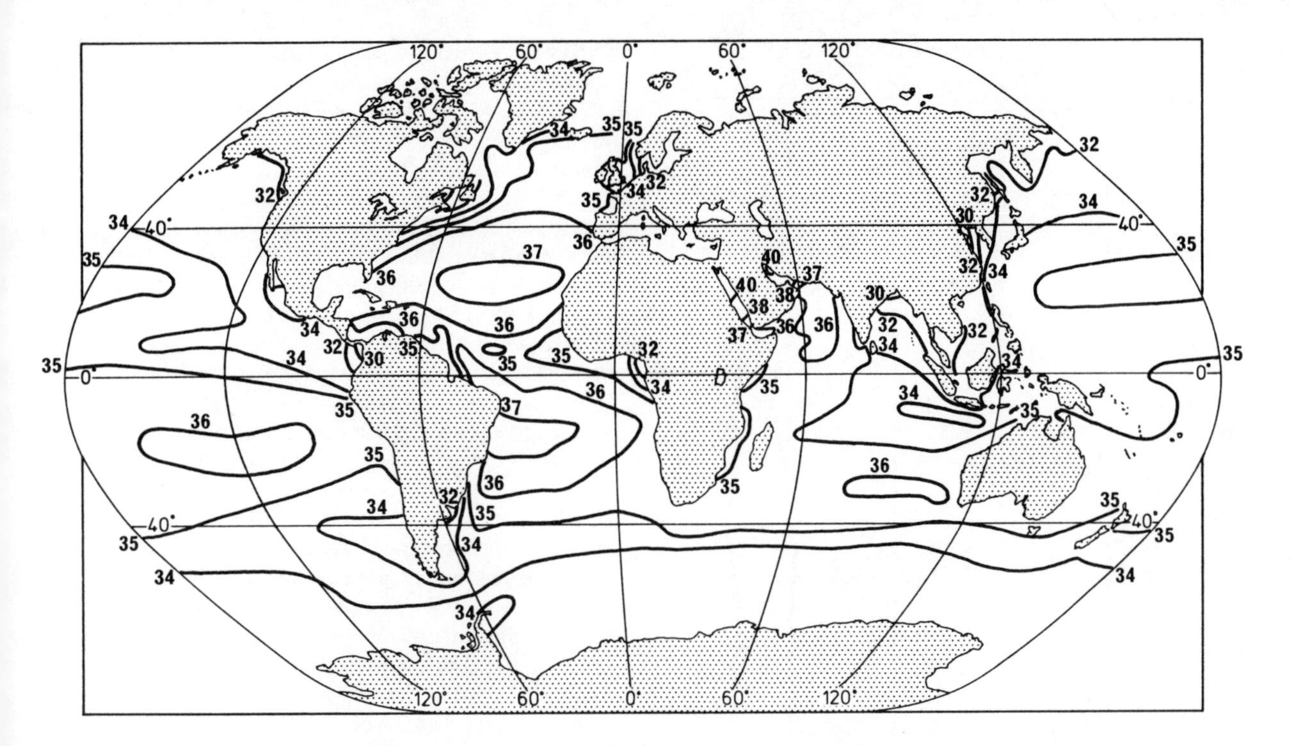

4. 2 Surface salinity of the oceans (‰) in the northern summer.

ceed further along the food chain. At any point in it they may be released back into the water when an organism dies and is decomposed, but this usually involves further sinking below the photic zone. Hence there is a net downward transport of nutrients in the ocean between uptake in phytoplankton and subsequent return to the dissolved state, and the effect of this can be seen in the mean distribution of nutrient salts with depth in the Pacific Ocean (Fig. 4. 3). For phytoplankton growth to continue uninhibited in the photic zone, nutrients must be transported back upwards to it; hence the importance of regions of upwelling to the fertility of the oceans.

In shelf seas in temperate latitudes, the concentration of nutrients varies with season rather than with depth (Fig. 4. 4). In spring, with increasing sunlight, a considerable growth of phytoplankton takes place, the 'spring phytoplankton bloom'. This may use up the stock of one or more essential nutrient salts almost completely, and so bring the phytoplankton bloom to an end. Sometimes sufficient nutrients re-accumulate for a secondary bloom later in the summer when the turbidity of the water has perhaps decreased, so allowing the sunlight to penetrate further. During the winter, however, phytoplankton growth comes virtually to a standstill, and thorough mixing of the water column re-distributes the nutrient salts which have been returned to the water by organic decomposition evenly from surface to bottom, ready for the bloom in the following spring.

DISSOLVED GASES

As well as salts, gases from the atmosphere are dissolved in sea water. There are two, oxygen and carbon dioxide, which are of special interest because they participate in biological and geochemical processes occurring in the ocean. Oxygen is released into the ocean by the process of photosynthesis (see Equ. 4. I), but conversely it is utilised by organic respiration and decomposition in the break-down of carbohydrates. Its concentration is therefore highest in the photic zone where it is provided by photosynthesis as well as by solution from the atmosphere, and lowest in water which has been below the photic zone for the greatest length of time and in which organic processes have been active. A compensation depth can be defined towards the bottom of the photic zone at which oxygen production just

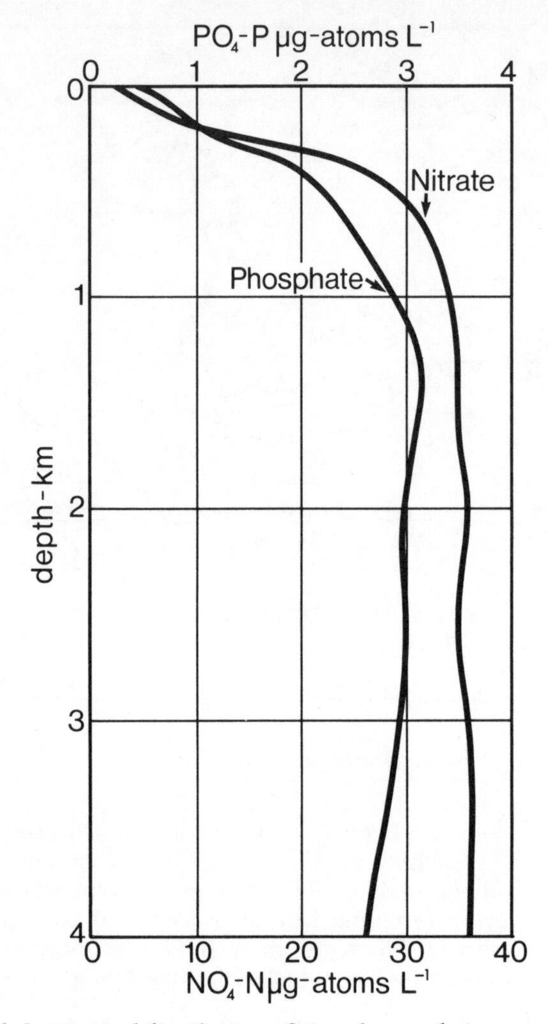

4. 3 Vertical distribution of phosphate and nitrate in the Pacific Ocean. The concentrations are given in terms of the weight in microgrammes of the nitrogen or phosphorus atoms per litre. (After H. U. Sverdrup et al.)

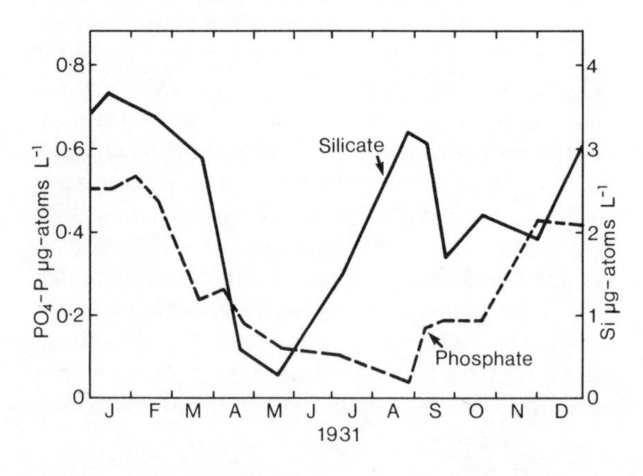

balances oxygen utilisation. At the ocean surface the dissolved oxygen concentration is usually at, or very close to, its saturation level, but since this increases as temperature decreases it varies from about 4.5 ml per litre in tropical latitudes to more than 8 ml per litre in polar seas. Between the surface and the compensation depth it might exceed its saturation value, but below this it decreases, though an actual oxygen profile with depth (Fig. 4. 5) is typically fairly complex, being closely dependent upon the location at which the water at each level was last at the surface and upon its subsequent movements.

Carbon dioxide plays the reverse rôle to oxygen in biological processes. It is utilised in photosynthesis (Equ. 4. I), and released by respiration. In addition to these factors and to the solution of atmospheric carbon dioxide in the ocean, we must also take account of the *carbonate system* to understand the distribution of CO_2 in the ocean. Carbon dioxide reacts with water to form carbonic acid. This, and the carbonate ions brought into the ocean by rivers which have flowed for example over limestone ($CaCO_3$), enter into a chain of reactions, namely

$$CO_2 + H_2O \rightleftharpoons H_2CO_3 \rightleftharpoons H^+ + HCO_3^- \rightleftharpoons 2H^+ + CO_3^{--} . \qquad (4. II)$$

All these reactions can proceed in either direction in the ocean, moving towards a state of equilibrium, though this is rarely achieved. Supersaturation of the water with respect to the carbonate ion tends to result in the removal of calcium carbonate by organisms whose remains subsequently settle to the ocean floor; undersaturation tends to lead to the dissolution of such sediments, and the increase in the solubility of calcium carbonate with pressure thus leads to a marked drop in the calcium carbonate content of sediments, particularly below about 4 km depth. The oceans therefore provide a great reservoir for carbon dioxide which, backed up by the carbonate sediments on the ocean floor, very effectively buffers the carbon dioxide content of the atmosphere from change.

4. 4 Seasonal variation of phosphate and silicate in the surface layer (0–25 m) of the English Channel during 1931. (After H. U. Sverdrup et al.)

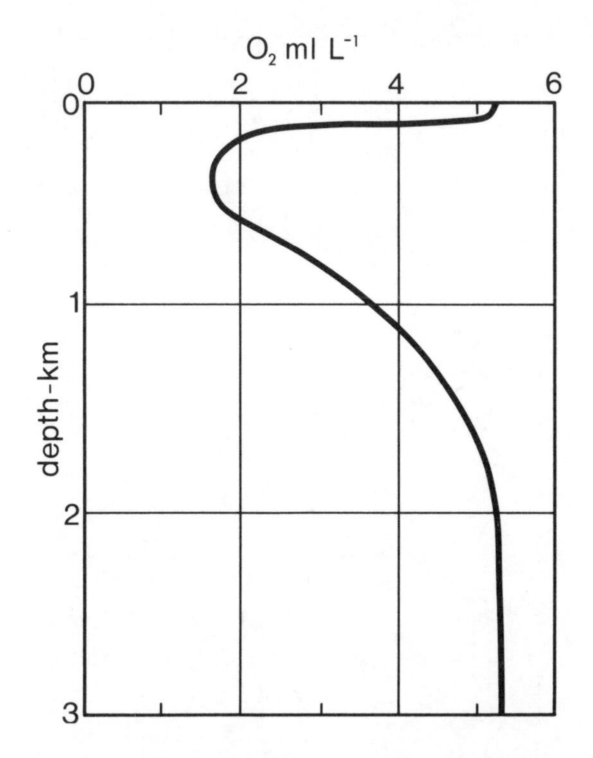

4. 5 Vertical distribution of dissolved oxygen in the tropical Atlantic Ocean. (After G. Dietrich)

Ice found in the oceans may have formed there by the freezing of sea water or the accumulation of snow, or it may have been transported there by glaciers or partially frozen rivers. The ice which forms on sea water is known as *sea-ice*. We have already noted that the presence of salt in water lowers its freezing point, and that when the salinity of sea water exceeds 25‰ its density increases as it is cooled right down to freezing point (Fig. 2. 3). Thus, unless a column of sea water is stratified by a higher salinity at greater depths, surface cooling leads to convection from surface to bottom, and a very considerable heat loss will be required in all but very shallow seas for ice to form. When sea water first begins to freeze, only water molecules coalesce to form ice crystals. The salt remaining behind adds to the salinity of the surrounding water, thus lowering its freezing point further, raising its density and so promoting more mixing. If freezing continues, these crystals (which are known as *frazil ice* and which give the water an oily appearance) grow and increase in number to cover the surface with a slush. As this thickens, it breaks up into pieces which are rounded by continual bumping against each other and which are known as *pancake ice*. During this process of growth, brine becomes trapped in enclosed spaces in the ice, so incorporating salt into sea-ice. If the temperature falls sufficiently, various salts will crystallize out from this brine, e.g. sodium sulphate at −8.2°C, but even at temperatures as low as −30°C most of the salt present in sea-ice is in a very concentrated solution trapped in the crystal structure. How much salt is so trapped in sea-ice depends on the speed with which the ice is formed, and thus on the air temperature during freezing. For example, the salinity of new ice formed from sea water of salinity 30‰ with air temperatures of −16°C was found to be about 5.6‰, while that formed from the same water but with air temperatures of −40°C was more than 10‰.

When sea-ice forms a more or less continuous layer, thicker than about 5 cm, it becomes *young ice*, and if this thickness exceeds 20 cm but the ice is still less than a year old it is called *winter ice*. *Polar ice*, which is more than a year old, is of lower salinity than winter ice because the salt, being mainly dissolved in brine, is able to leach out, particularly in summer when the cells in which it is trapped are opened up by melting. The salinity of old ice is also lowered by precipitation – as the ice thickens, its poor thermal conductivity reduces the heat loss from the water below; thus the rate at which sea water freezes to the base of the ice is reduced when fresh snow is added at the top.

Winds and waves break up sea-ice into fragments of varying size, carry it around the polar seas, and pile it up into various shapes and forms – hummocks, floes, fields, fast-ice, etc. The term *pack-ice* is applied to the entire floating ice fields in each hemisphere (Fig. 4. 6-b). In the Antarctic it consists mainly of large ice-floes which are relatively undistorted by lateral pressures, whilst in the Arctic it is much less even, with hummocks and ridges where ice has been piled up, and open strips and lanes (particularly along the coast) where it has been pushed apart. Pack-ice may be navigable where it is thin enough to be broken by specially strengthened vessels, or where the ice covers less than 6/8 of the surface and there is open water between. The signs which indicate that an observer is approaching an area of pack-ice include 'ice-blink' (a white gleam above the

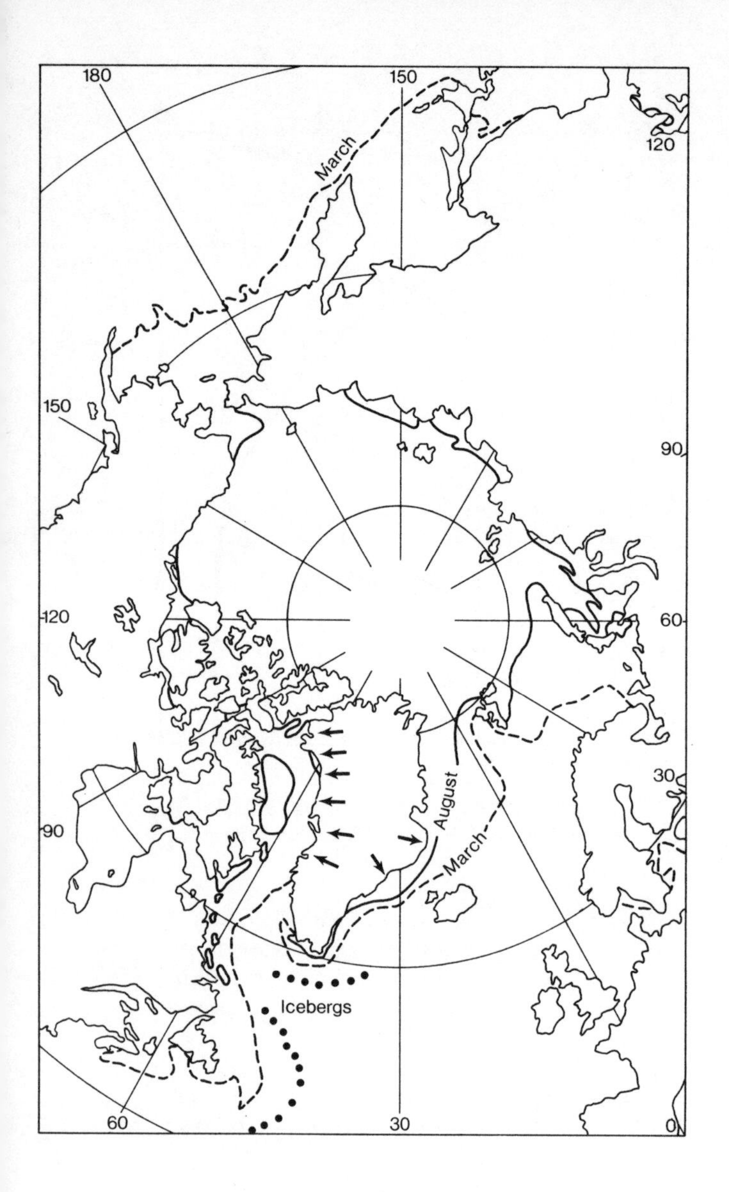

4. 6-a Average boundaries of sea-ice in August and March, and sources and probable limits of icebergs in the northern seas.

horizon resulting from the reflection of light at the ice surface), a decrease in swell which is damped out by the ice, a fall in sea-water temperature, and the appearance of certain birds. In due course the pack-ice will melt, some during the summer months in the polar seas and some, perhaps more than three years old, after it has been transported to lower latitudes by winds and ocean currents. When it melts it will dilute the surface water, and thus the salinity of the surface water in high latitudes is (particularly in the northern hemisphere) appreciably less than that in ice-free latitudes – mostly 30-33‰ against 35‰. This does, however, provide stratification and assist the formation of fresh ice in the winter.

Ice which is brought to the sea by glaciers forms *icebergs*. In the northern hemisphere most icebergs come from the valley glaciers of Greenland. Other sources include Spitsbergen, Franz Josef Land and Novaya Zemlya (Fig. 4. 6-a). Most of these discharge into the sub-arctic seas rather than into the almost enclosed Arctic Ocean. In the southern hemisphere, icebergs originate mainly from the shelf-ice of the Ross and Weddell Seas which spreads on to these seas, masking the coastline and resting on the sea floor until it is in sufficient depth of water to float. In the northern hemisphere there are thick ice sheets fringing parts of the Arctic Ocean (e.g. along the north coast of Ellesmere Island) which can be considered intermediate between sea-ice and land-ice. Although located over the sea, they are formed mainly from snow which has been consolidated and resembles that in glaciers. Large areas with thicknesses of perhaps 50 m, which break off occasionally and drift around the Arctic Ocean for many years, are known as *ice islands*, and a number of bases have been set up on them to make scientific observations. Ice which is carried down to the Arctic ocean by rivers, mainly from northern Russia in spring, melts rapidly and is of relatively little significance.

There is a marked difference between the icebergs of the North and of the South due to their differing origins. Arctic icebergs are relatively small but irregular in shape, and so, although they are usually denser than Antarctic ones (about 0.9×10^3 kg m^{-3}), the ratio of their height above the surface to their depth below is often larger. Shortly after calving, many are more than 60 m high and perhaps 250 m deep, but few are more than 1 km long. As they weather and melt, their height is reduced and substantial parts may break off, leaving the iceberg unstable so that it capsizes.

Antarctic icebergs are usually of a tabular form,

4.6-b Average pack-ice boundaries at the end of the winter and summer, and average northern boundary for icebergs, in the Antarctic Ocean. (After P. Groen)

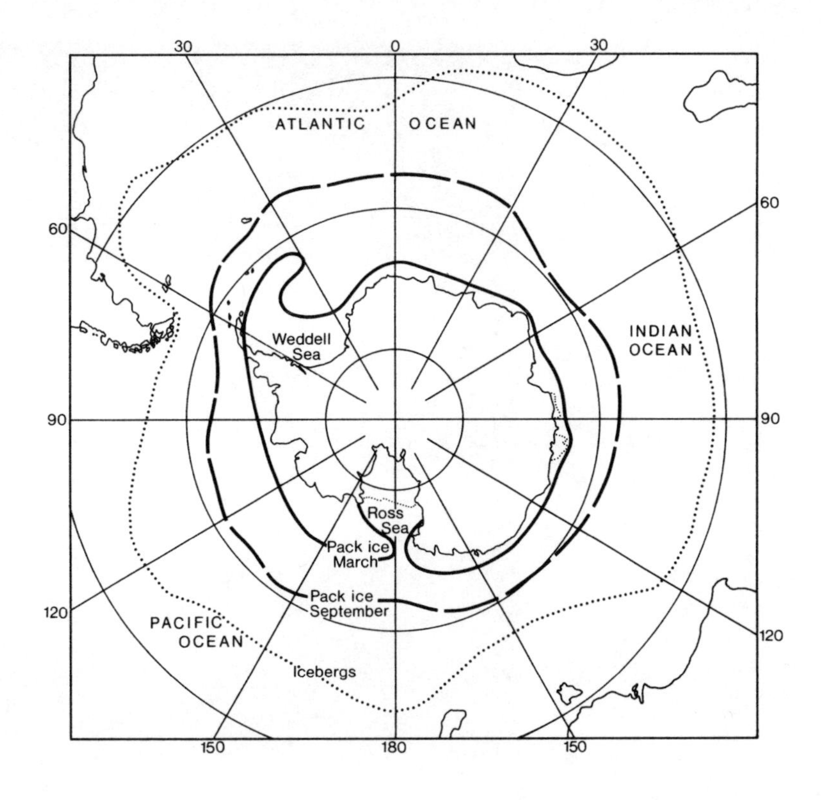

and often have surface areas of many square kilometres, but rarely exceed about 35 m in height. Their air content is usually quite high, because the shelf-ice from which they form has not been subject to as much pressure as the valley glaciers of the northern hemisphere, and therefore they are whiter in colour and less dense. Arctic bergs usually contain more soil and debris derived from erosion, and this further accentuates these differences in colour and density. Due to their greater size, icebergs usually travel over much longer distances than sea-ice before they melt, and their greater draught means that they are more influenced by ocean currents than by wind. Even after melting and breaking up into pieces the size of houses ('bergy bits') or of less than 10 m in length ('growlers'), they can still cause considerable damage to a ship colliding with them. In the western North Atlantic where they move into shipping lanes (the *Titanic* sank in 1912 after collision with an iceberg), an International Ice-Patrol monitors the positions and predicts the movements of ice which is a potential danger to shipping. The movements of icebergs will be discussed further in Chapter 14 when we consider the circulation of the oceans.

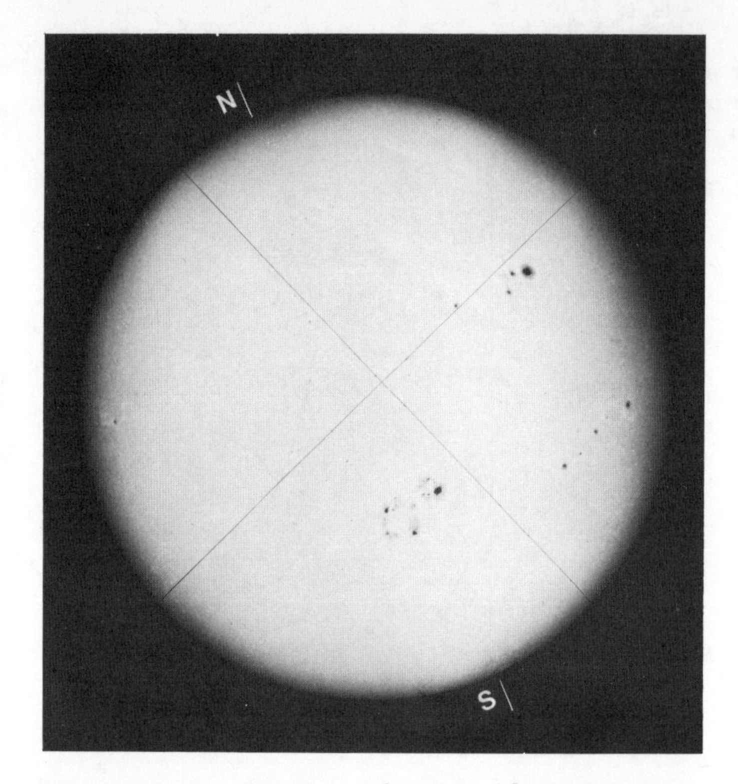

5.0 Above: the solar disc, with groups of sunspots – comparatively cool regions associated with strong, local magnetic disturbances.
Below: large eruptive prominence on the Sun. The white circle represents the Earth on the same scale. Such prominences shoot outwards at up to 100 km s^{-1} and dissipate themselves within hours. Both phenomena affect the solar radiation which reaches the Earth.

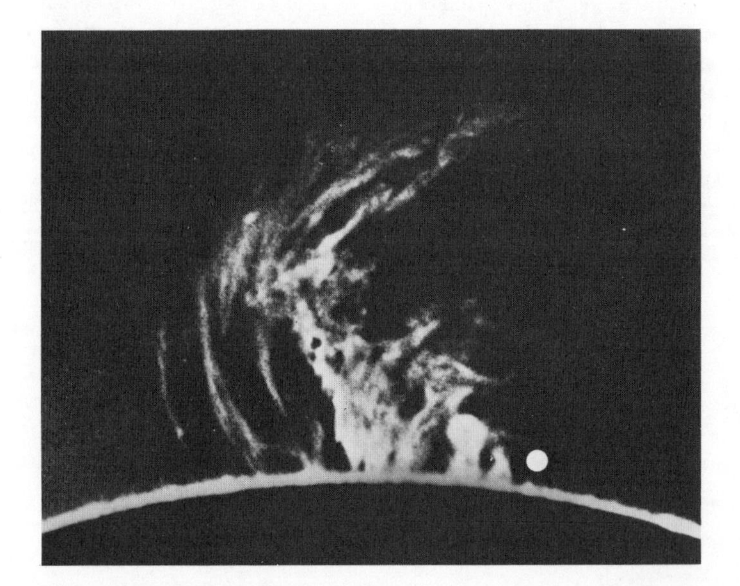

5 The Energy Source and Heat Distribution

Solar radiation □ Transmission through the atmosphere □ Effect of the Earth's surface □ Albedo □ Heat balance and back radiation □ Heat transfer within and between the atmosphere and ocean □ Diurnal and seasonal variations of temperature □ Global distribution of air and sea surface temperature.

IN GENERAL WE MAY ASSUME that the sole source of thermal energy for the atmosphere and ocean is the Sun. There is some conduction of heat outwards from the interior of the Earth and some frictional dissipation of tidal energy, but on average the amount of heat received from the Sun exceeds these by factors of about 10^4 and 10^5 respectively. We shall consider the nature of solar radiation before examining the fate of that part of it which reaches the planet Earth.

SOLAR RADIATION

The Sun can be taken for our purposes to act as a 'black body' with a surface temperature of 6 000 K. It emits electro-magnetic radiation, a form of energy which travels with a wavelike motion through space at a speed of 3×10^8 m s^{-1} (the speed of light). The total amount of radiation which it emits per unit surface area (E) is given by Stefan's Law, $E = \sigma T^4$, where σ is a constant and T the absolute temperature of the surface. This energy is spread through a wide range of wavelengths as shown in Figure 5. 1. This energy spectrum can be conveniently divided into three sections, viz:

>0.7 μm wavelength – infra-red radiation accounting for about 48% of the total,

0.4 – 0.7 μm – radiation visible to the human eye – about 43% of the total,

<0.4 μm – ultra-violet radiation and X-rays – about 9% of the total.

Approximately 99% of the Sun's radiation is of wavelengths between 0.15 μm and 4.0 μm, and the maximum emission is at a wavelength of about 0.5 μm (blue-green light).

The Earth follows an elliptical path around the Sun, its mean distance away being about 150

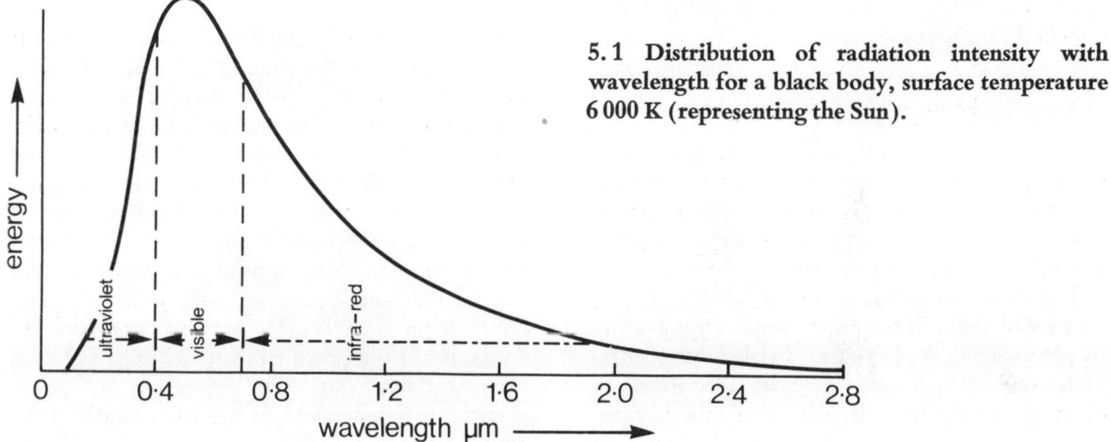

5.1 Distribution of radiation intensity with wavelength for a black body, surface temperature 6 000 K (representing the Sun).

million km, but this varies at the present time by about 5 million km in the course of a year. A surface exposed perpendicularly to the Sun's rays at the mean distance of the Earth from the Sun, and with no atmosphere above it, would receive energy of ca. 1.36 kW m^{-2}. This is known as the *solar constant*, but whether or not it *is* constant is open to doubt. Until recently it had to be determined from measurements of the radiation received at ground level, with an allowance being made for the reduction of radiation as it passed through the atmosphere. The variations thus observed this century have been within the limits of error of the method ($\pm 5\%$). The use of space probes should ultimately enable us to determine whether there are variations in the solar constant – perhaps associated with sunspot activity, the visual indication of solar activity – which might be sufficient to bring about changes in the Earth's climate. Problems in maintaining the calibration of instruments mounted on Earth satellites have so far prevented sufficiently reliable data being obtained.

If there were no atmosphere around the Earth, the amount of radiation received by unit area of the Earth's surface in a day would depend upon the length of time the area was exposed to the Sun's rays, the angle between the Sun's rays and the Earth's surface, and the distance of the Earth from the Sun. These factors vary with latitude and season, and their effect is shown in Figure 5. 2. Note that the summer maxima of radiation which would be received in the absence of an atmosphere increase with increasing latitude, but because the winter values decrease more rapidly, the annual mean values decrease with increasing latitude. Because perihelion (the nearest approach of the Earth to the Sun) is reached in the southern summer, more radiation is received then than in the northern summer.

TRANSMISSION THROUGH THE ATMOSPHERE

Figure 5. 2 does not represent the amount of radiation actually received at the Earth's surface because, as it passes through the atmosphere, radiation is subject to absorption and scattering, and also to reflection by clouds.

The radiation in the shortest wavelengths is *absorbed* by gases in the upper atmosphere leading to photochemical reactions. In absorbing ultra-violet and X-rays the molecules and atoms of these gases may lose electrons and thus become positively charged ions. The region of the atmosphere in which the concentration of ions and electrons is greatest (60 to 300 km above the Earth's surface) is the *ionosphere*, and permits long-distance transmission of radio waves by reflecting them back to the surface. Alternatively, molecules may be dissociated into single atoms by ultra-violet radiation. Oxygen is readily dissociated in this way, and single atoms of oxygen then combine with oxygen molecules (O_2) to form molecules of ozone (O_3). Ozone is in turn broken up by the absorption of slightly longer wavelength ultra-violet radiation, or it can be destroyed by collision with another atom of oxygen, when two oxygen molecules result. Although ozone is formed mainly above 40 km, it is found in its greatest concentration between 20 and 35 km. This is due to downward transport in the atmosphere to a level where it is not destroyed so rapidly by incoming radiation. Here it completes the absorption of ultra-violet radiation which would be harmful to life, and also absorbs a small amount of radiation in some longer-wavelength bands. Ozone is itself poisonous in more than very minute concentrations, but it is virtually absent in the atmosphere below 10 km, being destroyed in the oxidation of substances from the Earth's surface. In all, some 3% of the incoming radiation is absorbed by gases above 10 km in the atmosphere, principally ozone.

In the lower atmosphere the only gaseous constituent able to absorb significant amounts of the remaining solar radiation is water vapour. Typically some 10% of the total solar radiation is absorbed by it, but the amount varies considerably according to its local concentration. Clouds and dust particles also absorb some of the incident solar radiation.

When electro-magnetic radiation encounters small particles in the atmosphere and is not absorbed, it is *scattered*. If the particles have a radius less than one-tenth the wavelength of the radiation, the intensity of scattering is greatest at the shortest wavelengths. Thus, of the radiation visible to the human eye, blue light is scattered most by air molecules, giving the sky its characteristic colour. Near sunrise and sunset the sky around the Sun appears red and yellow because these longer wavelength components remain predominant in the beam from which blue light has been scattered out. The scattering by larger particles is independent of the wavelength of the

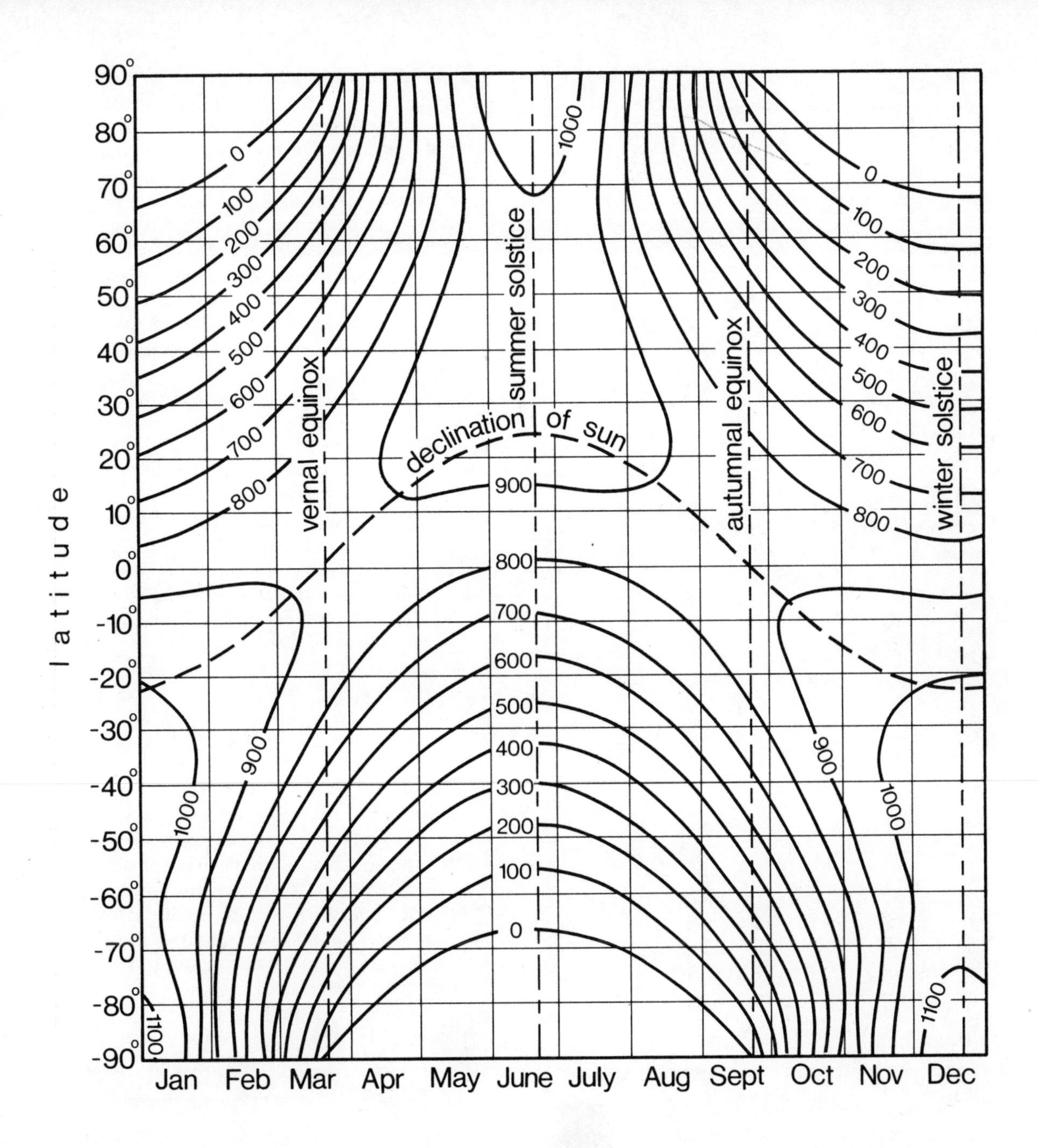

5.2 Daily totals of solar radiation in cal cm^{-2} which would be received at the Earth's surface if there were no atmosphere present, as a function of latitude and season. (From Smithsonian Meteorological Tables)

radiation, and thus leads to a general whitening of the sky, causing fog or haze. Some of the radiation which is scattered is subsequently absorbed in the atmosphere, and some eventually reaches the Earth's surface (perhaps after multiple scattering), but about 7% of the incident radiation is lost to space by scattering.

Reflection occurs in the atmosphere when solar radiation is incident on clouds. The proportion which the cloud reflects, its *albedo*, depends on the

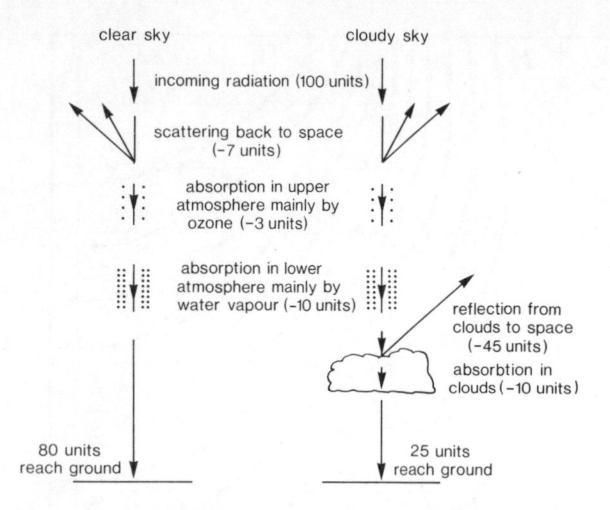

5. 3 Reduction of solar radiation intensity as it is transmitted through the atmosphere. (The values given are very approximate, and are intended to represent typical conditions.)

cloud type and thickness; it may be as much as 80% for a thick stratocumulus cover, but averages about 55%, and most of this is lost directly to space.

The effects on the incident radiation of absorption, scattering and reflection are summarised in Figure 5. 3. The amount of solar radiation which reaches the Earth's surface may be as much as 80% of that incident at the top of the atmosphere with a clear sky, or be less than 20% with an overcast sky. Cloud cover apart, variations are brought about by the amounts of water vapour and dust in the atmosphere, and by the length of the radiation's path through the atmosphere which is essentially a function of the Sun's elevation in the sky.

EFFECT OF THE EARTH'S SURFACE

The radiation which reaches the Earth's surface may be absorbed there, be transmitted downwards if it encounters a material which is transparent to it, or be reflected. The albedo of the surface depends on its substance and texture, the angle of incidence of the radiation, and the wavelength of the radiation. Table 5. 1 gives total albedos (i.e. for solar radiation as a whole) with the Sun directly overhead for a variety of surfaces. Most land surfaces have values between 10 and 30%. For a calm water surface, while the Sun is within 50° of the zenith, it is less than 5%, but when the Sun is low in the sky this increases dramatically – over 50% when the Sun is 80° from the zenith. The mean albedo for the entire planet Earth, which includes the clouds in the atmosphere, is about 35%.

The only part of the Earth's surface which is significantly transparent to solar radiation is water. Radiation reaching a water surface which is not reflected penetrates downwards, being refracted towards the vertical as it does so. It is then subject to absorption and scattering, mainly by suspended particles. In filtered sea water about 40% of the incident radiation reaches 1 m depth and 22% reaches 10 m, whereas in average oceanic water the corresponding figures are 35% and 10%, and in turbid coastal water they are 23% and 0.5%. The rates at which different wavelengths of solar radiation are absorbed and scattered vary considerably. Infra-red and red light penetrate to the least depth; in clear ocean water blue light penetrates furthest, whereas in turbid coastal waters it is green and yellow light.

The absorption of radiation, either at the solid Earth's surface or within the oceans, leads to heating. The heat may be transmitted downwards by conduction or, in the case of fluids, by convection. The process of conduction is relatively slow and, where the Earth's surface is composed of solid material, diurnal heating is not discernible more than about 0.5 m below the surface. In the ocean convection is the major process leading to the downward transmission of heat. Apart from thermal convection (which results from surface cooling in the ocean), vertical mixing is induced by wind waves and turbulent currents, and diurnal heating is typically spread through a depth of 10 m or more in the open ocean.

The absorption of heat leads to an increase in temperature, the magnitude of the increase being

Table 5. 1: Some typical albedos

SURFACE	ALBEDO
Fresh snow	up to 90%
Thawing snow	about 40%
Desert sands	35%
Green grass or forest	10–25%
Dry ploughed field	12–20%
Cities and rocks	12–18%
Moist soil	about 10%
Calm water	2%

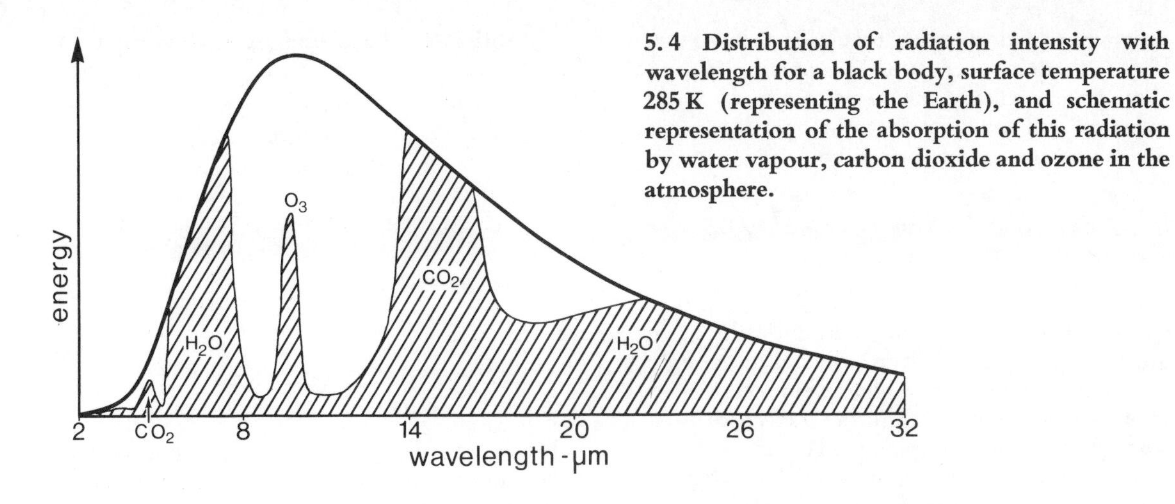

5.4 Distribution of radiation intensity with wavelength for a black body, surface temperature 285 K (representing the Earth), and schematic representation of the absorption of this radiation by water vapour, carbon dioxide and ozone in the atmosphere.

inversely proportional to the specific heat of the substance, or possibly to a change of state in the case of water. The specific heat of water is some five times that of rock or dry soil, but because water is less dense its heat capacity is only about twice that of an equal volume of rock. Thus if equal volumes of water and rock receive a given input of heat, even if no water evaporates and the heat is distributed evenly, the increase in temperature of the water will be only about half that of the rock, and the difference would be greater if dry soil were considered instead of rock.

HEAT BALANCE AND BACK RADIATION

If the Earth continued to absorb solar radiation without any loss of heat, its temperature would rise indefinitely. This does not happen because the Earth emits electro-magnetic radiation into space. Taking mean annual values, and ignoring any change in the Earth's mean annual temperature from one year to the next, a balance must exist between incoming solar radiation and outgoing terrestrial radiation.

The Earth's temperature controls both the amount of radiation which is emitted and the energy spectrum of this radiation. From Stefan's Law the average 'black-body' temperature for the Earth required to achieve this balance is found to be 250 K (−23°C). This is known as the *planetary temperature* of the Earth. It is much lower than the average temperature of the Earth's surface because a substantial amount of the radiation emitted by the Earth's surface is either absorbed or reflected back in the atmosphere.

From Figure 5. 4 it can be seen that the wavelength at which the maximum emission of radiation takes place from a black body with a surface temperature of 285 K, close to the Earth's mean surface temperature, is about 10 μm. Almost all of the radiation is of wavelengths greater than 4 μm, which is thus taken as the value separating 'short-wave' solar radiation from 'long-wave' terrestrial radiation. The gases in the atmosphere which are particularly effective in absorbing long-wave terrestrial radiation are water vapour, carbon dioxide and ozone. They absorb virtually all terrestrial radiation of wavelengths less than 8 μm and more than 12 μm, but between these values they leave a 'radiation window' through which radiation is lost to space when the sky is clear. Clouds both absorb and reflect long-wave radiation. The atmospheric constituents which absorb terrestrial radiation in turn emit long-wave radiation in all directions, some out into space but some back to the Earth's surface. They therefore act as a layer of insulation around the Earth analogous to the glass of a greenhouse, and their effect on Earth temperatures has been called the *greenhouse effect*. The amount of carbon dioxide in the atmosphere, which is believed to have increased by perhaps 10% in the past 70 years due to the burning of fossil fuels, has an important bearing on the magnitude of this greenhouse effect, and could be responsible for changes in atmospheric temperatures.

The balance between incoming and outgoing radiation is achieved by adjustments in the Earth's temperature. If there is a net gain in radiation the Earth's temperature rises, leading to an increase in outgoing radiation so that balance is restored at a higher Earth temperature. To obtain even an approximate balance, however, mean values of radiation and temperature must be taken for a whole year and for the entire Earth.

HEAT TRANSFER WITHIN AND BETWEEN THE ATMOSPHERE AND OCEAN

If annual mean values of the incoming and outgoing radiation at any one location on the Earth are determined, an imbalance will almost certainly be found. This is because there are processes other than radiation which lead to the transfer of heat, particularly in the atmosphere and ocean.

The majority of incoming radiation which is absorbed by the planet Earth is absorbed at the surface, either land or water, after being transmitted through the atmosphere. As well as heating the atmosphere by the emission of long-wave radiation, the Earth's surface heats the atmosphere by the conduction of heat across the interface between the water or land and the air, followed by convection within the atmosphere. The process of conduction alone would lead to very little heat transfer, but conduction is only required through a layer of the order of millimetres or less in height. Vertical air motion, which may result from thermal convection or from horizontal motion over a rough surface (*turbulent convection*), then transfers the heat on upwards, maintaining a sufficient temperature gradient at the interface for conduction to continue fairly rapidly.

Table 5. 2: Heat budget of the ocean

Net gain of heat:	*units*	*Net loss of heat:*	*units*
by short-wave radiation	100	by long-wave radiation to atmosphere & space	41
		by conduction of sensible heat to atmosphere	5
		by evaporation	54
			100

In the case of a water surface, there is another very important process whereby heat received from the Sun is transferred into the atmosphere. This is the process of evaporation from the surface followed by condensation in the atmosphere. For each gram of water which evaporates from the ocean some 2.47×10^3 Joule of heat are required which the water vapour carries into the atmosphere as latent heat, to be subsequently released when condensation occurs.

Attempts have been made to evaluate the heat budget of the ocean, and Table 5. 2 gives some indication of the relative importance of the three main ways in which heat is lost from the entire ocean between 70°N and 70°S. There are considerable differences from one location to another and from one season to another, but in general evaporation is most important and conduction of sensible heat followed by convection in the atmosphere is least important. To produce this pattern

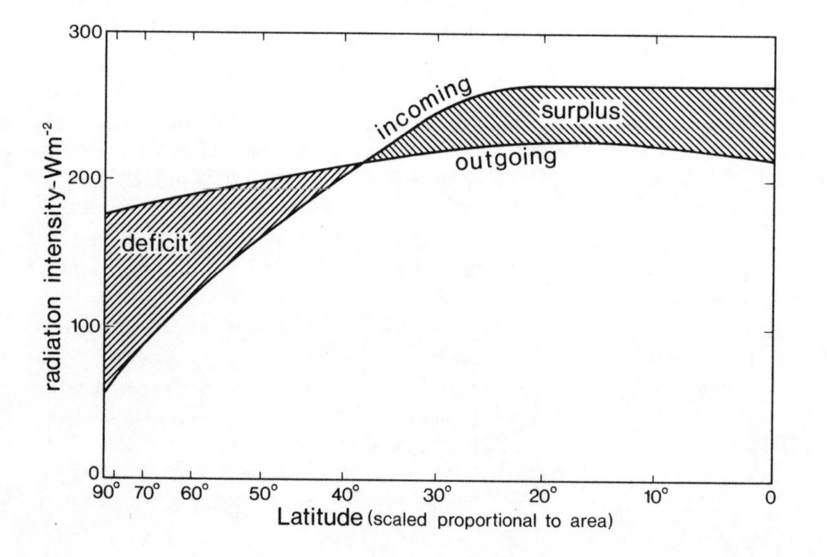

5. 5 Long-term mean values of incoming, short-wave radiation and long-wave, outgoing radiation for the Earth-atmosphere system, averaged over zones of latitude. (After J. C. Johnson)

the average ocean surface temperature must exceed that of the overlying air, and, more important, the vapour pressure of the air overlying the ocean must be less than the saturated vapour pressure of air at the temperature of the surface water so that evaporation takes place. There are, however, important exceptions to this general pattern. In the region of the Grand Banks off Newfoundland, for example, air temperatures exceed sea-surface temperatures in the spring. This leads to the transfer of heat from the atmosphere to the ocean so that condensation takes place on the sea surface and immediately above it, causing fog. This makes the water at the surface become warmer and thus less dense, while the overlying air becomes cooler and denser. Thus thermal convection is suppressed in both fluids, making the transfer of heat relatively slow except when there are strong winds.

Horizontal transfer (*advection*) of heat is necessary to compensate for a net loss of heat by radiation in high latitudes and a net gain by radiation in low latitudes (Fig. 5. 5). The change-over from a surplus to a deficit in the net annual radiation balance occurs at about 37° latitude N and S. Without any advection of heat, temperatures in equatorial latitudes would increase by more than 10°C, whereas in polar latitudes they would decrease by more than 20°C. This would considerably increase the proportion of the Earth's surface covered by ice and snow, which in turn would increase the surface albedo in middle and high latitudes, so leading to further cooling in these areas. Most estimates attribute more than 80% of this advection of heat as taking place in the atmosphere, where the global winds carry warm air and water vapour with its latent heat, polewards. An estimate using recent measurements from satellites of the Earth's radiation budget, however, shows the ocean contribution to average 40% of the total poleward energy transport between the equator and 70°N, and to account for 74% of it at 20°N. The winds and ocean currents not only correct the radiation imbalance between low and high latitudes, but themselves depend on the uneven distribution of heat over the Earth's surface for the energy which maintains them.

DIURNAL AND SEASONAL VARIATIONS OF TEMPERATURE

When periods of time shorter than a year are taken, radiation imbalances can be expected, leading to changes in temperature which can be predicted if certain simplifying assumptions are made.

In the case of the diurnal variation of temperature at the Earth's surface, if we assume that there is no air motion (either vertically or horizontally), that the sky remains clear, and that the surface is solid and there is no conduction of heat downwards, the variation of temperature with time (dT/dt) is proportional to incoming short-wave

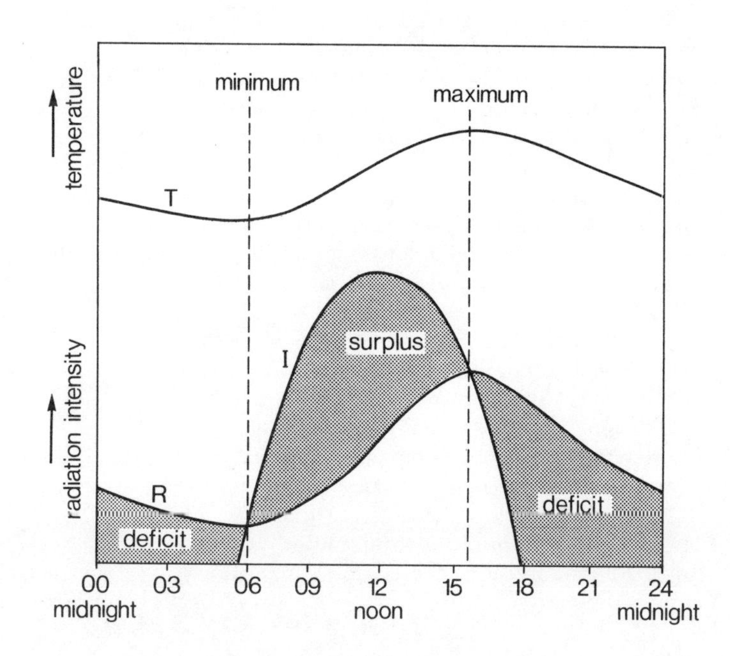

5. 6 Incoming short-wave radiation (I), outgoing long-wave radiation (R), and temperature (T) close to the ground during a diurnal cycle.

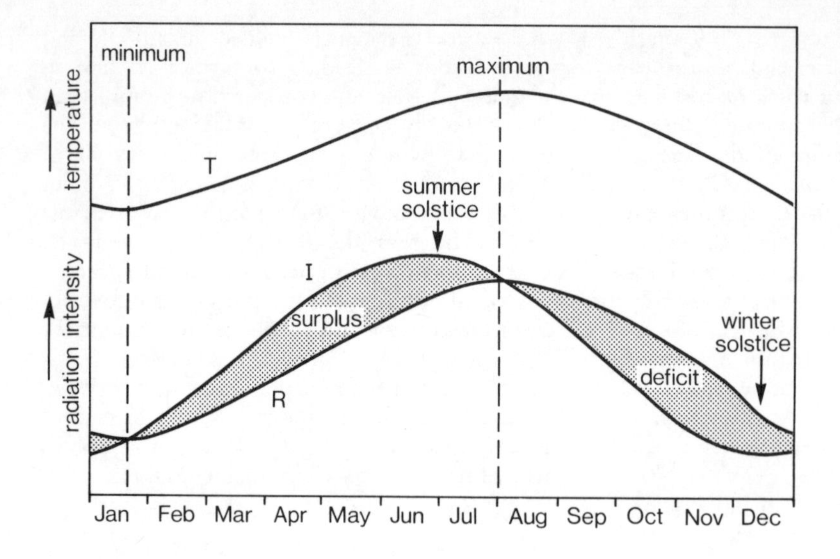

5. 7 Daily means of incoming short-wave radiation (I), outgoing long-wave radiation (R), and temperature (T) close to the ground during an annual cycle.

radiation (I) minus net outgoing long-wave radiation (R). I is a function of time, varying from zero between sunset and sunrise to a maximum at mid-day, and R is a function of temperature (T), related to it essentially by Stefan's Law. Figure 5. 6 shows how, if the value of T is known at midnight, a value of R can be determined which will lead to T decreasing as I is zero and thus $I-R$ must be negative. As T decreases, R will also decrease and thus the rate of change of temperature with time will decrease. At sunrise the value of I will start to increase, and about half an hour later it will equal the value of R, making (dT/dt) zero, and thus T will be at its minimum value. Subsequently $I-R$ is positive, and thus T increases and so also does R. However I remains greater than R until some three hours after mid-day when I has begun to decrease. At this time (dT/dt) again equals zero and T is at its maximum value. During the remainder of the day R is greater than I and thus both T and R decrease through to midnight.

Similar considerations can be used to predict the seasonal variation of temperature close to the ground (Fig. 5. 7). In this case the variation of I with time, using mean daily values of incoming short-wave radiation, can be represented by a sine wave, having its maximum value at the summer solstice and its minimum value at the winter solstice. The maximum and minimum temperatures are typically experienced about one month after the respective solstice.

The main features of the patterns described above can be seen in the variations of average air temperatures measured about one and a half metres above the ground at mid-continent locations – curves I in Figure 5. 8. At maritime locations (e.g. on an island or at a coastal situation) the pattern is less distinct, with a smaller range and the maxima and minima usually coming later in the day or year – curves II in Figure 5. 8. This results from:

(a) the greater penetration of heat downwards in water (due mainly to the processes of convection, but also to the transparency of water to solar radiation);

(b) the greater heat capacity of water compared to land;

(c) the large latent heats of fusion and evaporation of water – as well as being able to gain heat without any increase in temperature when evaporation takes place, water can lose heat without any decrease in temperature when it freezes.

The seasonal range of water temperatures decreases with increasing depth in the ocean (Fig. 5. 9), and at the bottom of the so-called *active layer*, typically about 200 m deep, variations associated directly with the incident radiation cycle can no longer be distinguished.

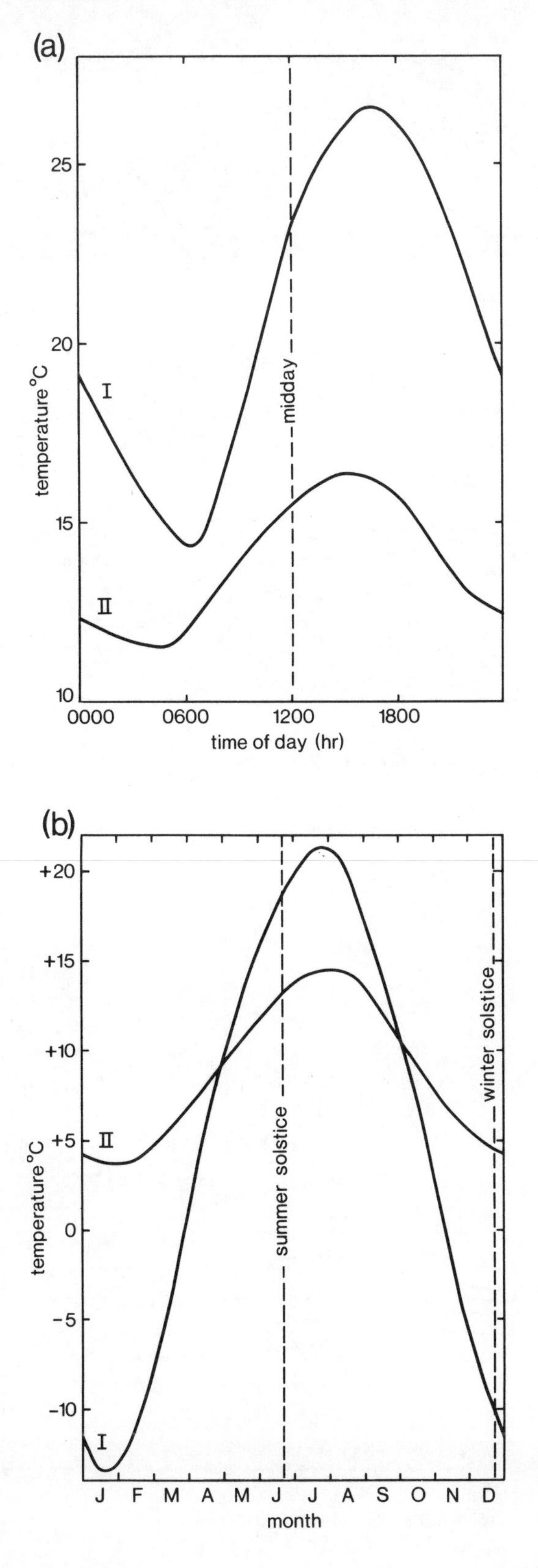

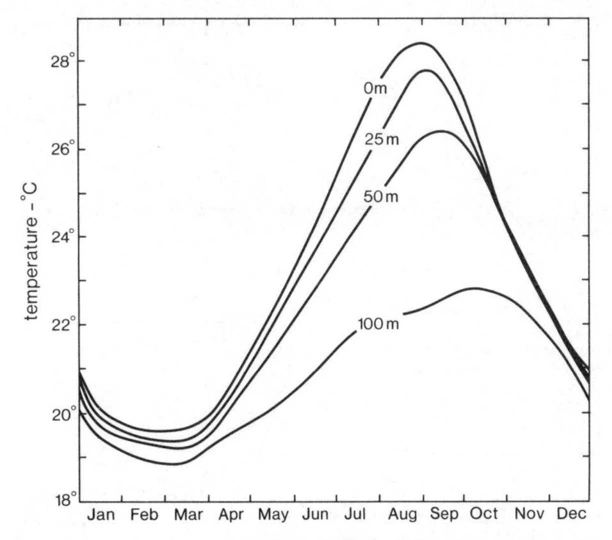

5.9 Annual variations of water temperature at different depths in the Kuroshio off the south coast of Japan. (After H. U. Sverdrup et al.)

HORIZONTAL DISTRIBUTIONS OF TEMPERATURE

Patterns of sea surface temperatures over the open ocean (Fig. 5.10) show values ranging from freezing point (about −1.9°C at a salinity of 35‰) to some 30°C. The seasonal range of temperature is widest in middle latitudes, and greater in the northern hemisphere than in the southern hemisphere, due essentially to the larger amount of land at middle latitudes in the northern hemisphere. The greatest seasonal contrast is in the North Pacific, where the mean range along 40°N is almost 10°C. Further, the mean position of the zone of maximum temperature, the *thermal equator*, is displaced to about 5°N. This, too, stems from the distribution of land and sea in the two hemispheres, and in particular from their effect on oceanic circulation and from the high albedo of the Antarctic continent. The paths of

---

5.8 (a) Diurnal variation in July, and (b) annual variation of air temperature at (I) Bismarck in the continental interior of North America (46° 48' N, 510 m above sea level), and (II) Fort William on the west coast of Scotland (56° 49' N, 50 m above sea level).

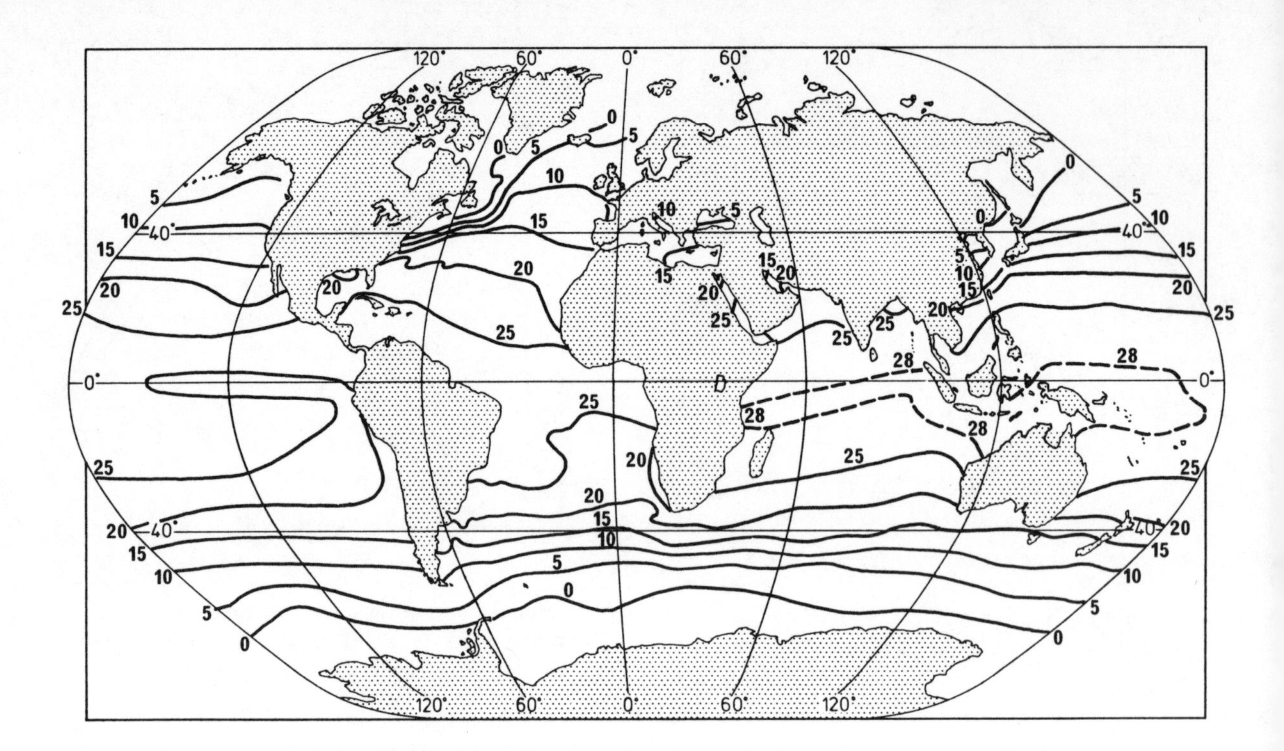

5. 10-a Surface temperature of the oceans (°C) in February.

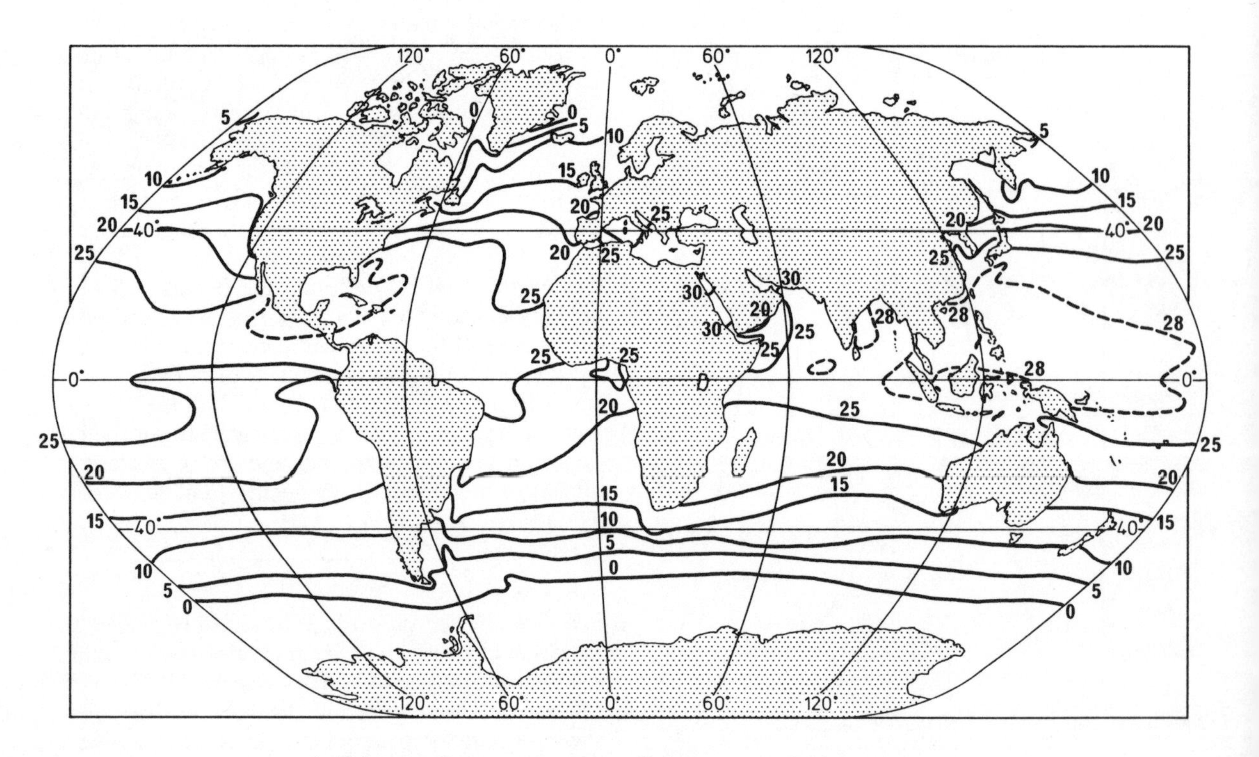

5. 10-b Surface temperature of the oceans (°C) in August.

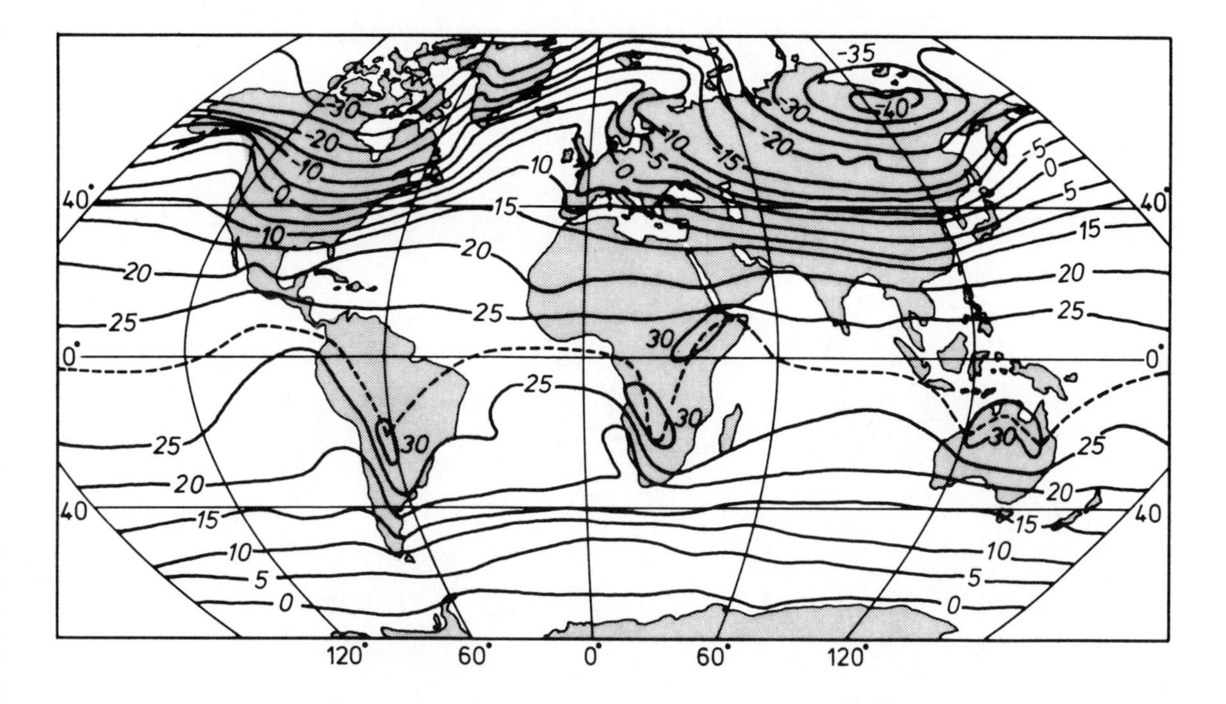

5. 11-a Air temperature reduced to sea level (°C) in January. The approximate position of the thermal equator is shown by the dashed line.

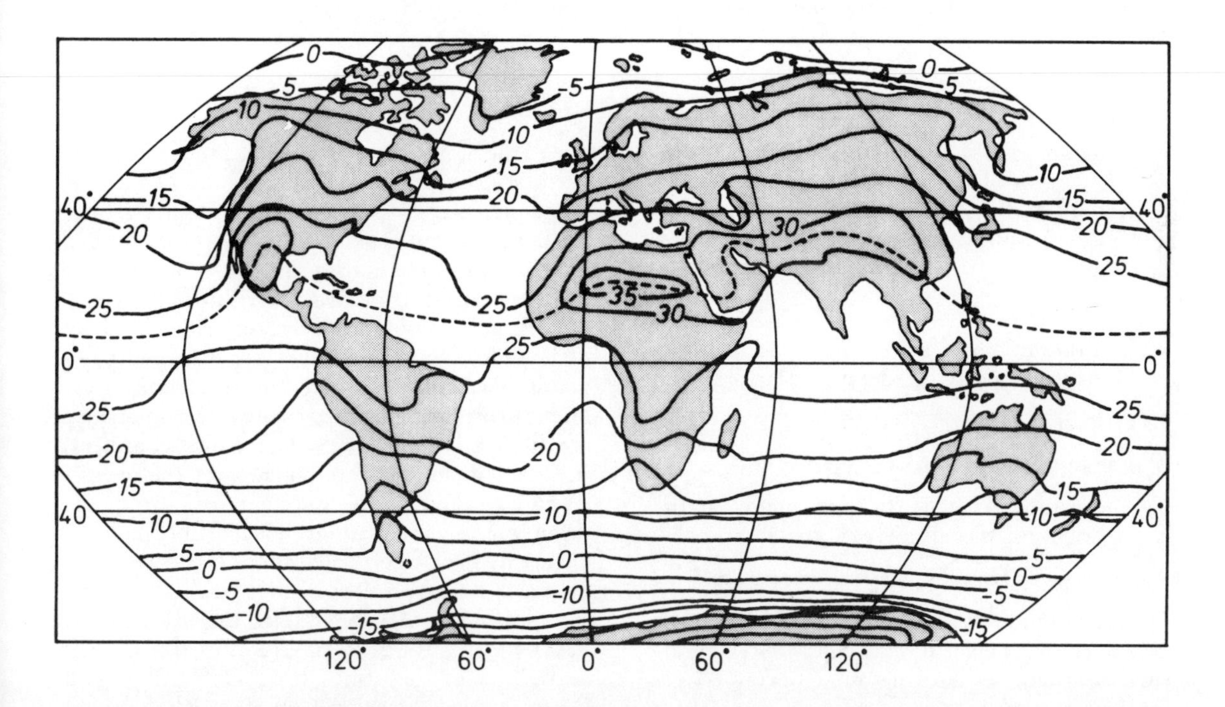

5. 11-b Air temperature reduced to sea level (°C) in July. The approximate position of the thermal equator is shown by the dashed line. (After R. G. Barry and R. J. Chorley)

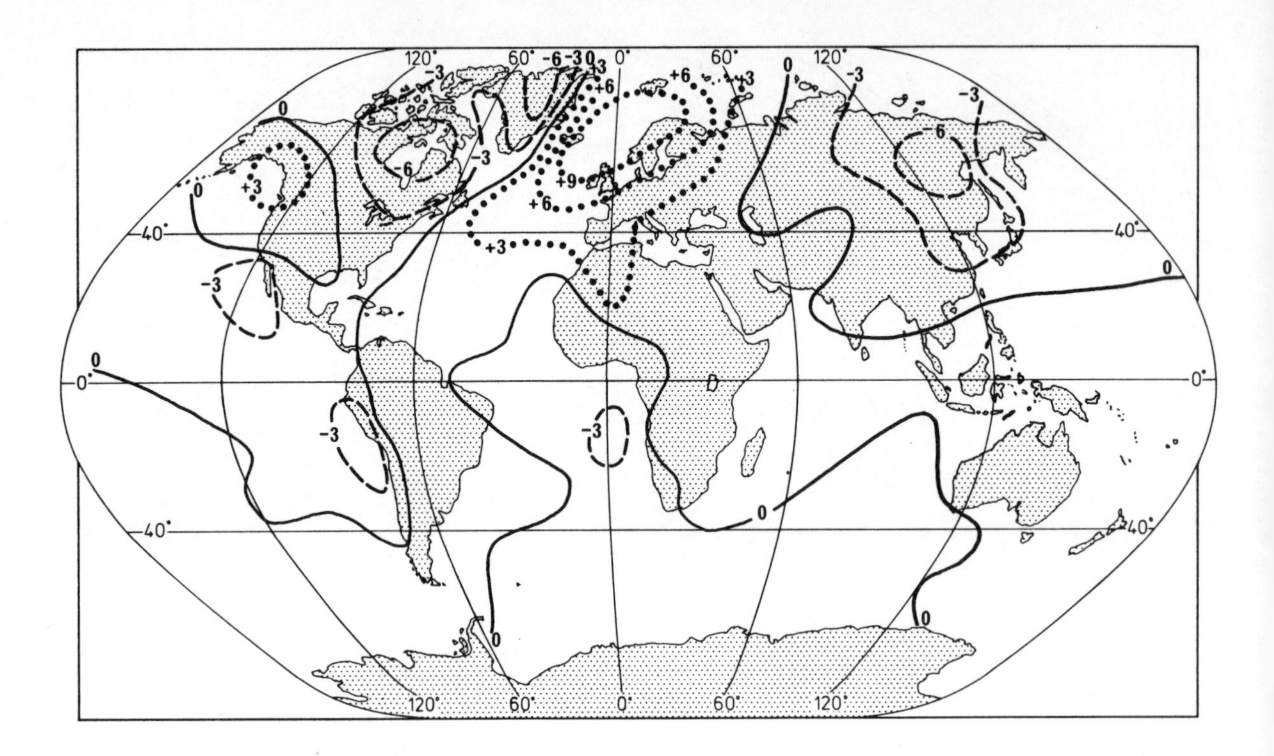

5. 12 Mean annual air temperature anomalies from the mean for that latitude (°C). Dotted lines indicate positive and dashed lines negative anomalies.

the ocean currents which contribute to the poleward advection of heat (e.g. the Gulf Stream in the western North Atlantic) stand out clearly by their marked positive temperature anomalies for their latitude.

The distribution of air temperature over the Earth's surface depends on four major factors, viz:

(1) latitude;

(2) the altitude of the surface;

(3) the nature of the surface, in particular the distribution of land and sea;

(4) advection of heat by winds and currents.

The effect of the altitude of the land surface will be considered in Chapter 6, and Figure 5. 11 eliminates this factor by showing air temperatures reduced to sea level.

The effect of latitude is for temperatures to decrease as latitude increases outside the tropics, due to the Sun's rays being inclined more from the vertical in higher latitudes. This means that the incoming solar radiation passes through a greater thickness of atmosphere and is spread over a greater area of the Earth's surface, so that a larger proportion of it is reflected at the Earth's surface.

Isotherms do not, however, run parallel to lines of latitude. In summer they are displaced polewards over continents (i.e. air temperatures are higher over land masses), whereas in winter they are displaced towards the equator over continents (i.e. air temperatures are higher over the ocean). Mean *annual* air temperature anomalies from the mean for a given latitude (Fig. 5. 12) indicate essentially the effects of advection. These anomalies reach positive values of more than 10°C over the North-East Atlantic due to the advection of heat by the Gulf Stream and North Atlantic Drift as well as in the atmosphere, contrasting with negative values of 4 to 5°C in the neighbourhood of Labrador, in similar latitudes on the opposite side of the North Atlantic.

6 Vertical Stability and Temperature Distribution

Stability of a fluid □ Stability in the oceans and atmosphere □ Adiabatic changes and potential temperature □ Assessment of stability: σ_t in the ocean, adiabatic lapse rates in the atmosphere □ Föhn winds □ Vertical temperature structure of the atmosphere: troposphere (average lapse rates and inversions), stratosphere, mesosphere and thermosphere □ Vertical temperature structure of the ocean and lakes □ Thermoclines.

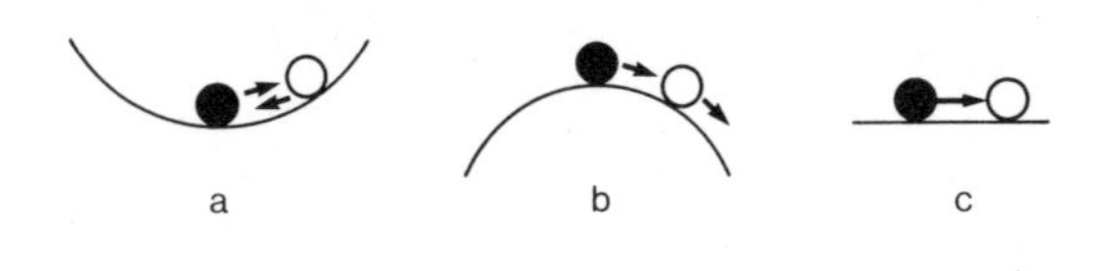

6.1 Examples of (a) stability, (b) instability, and (c) neutral stability.

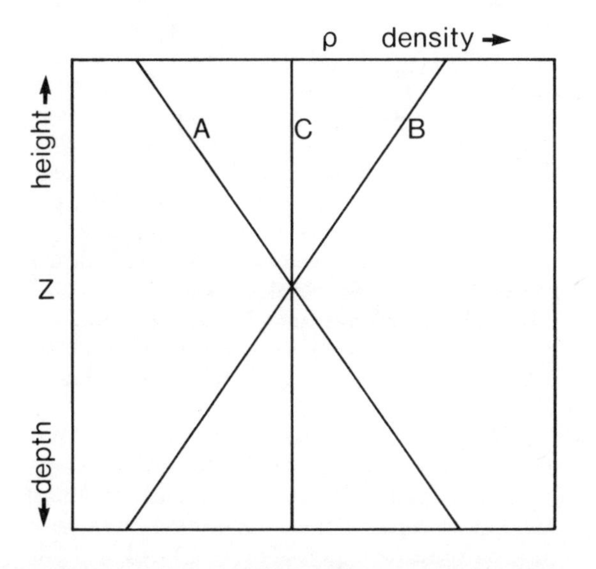

6.2 Variation of density with depth (or height) in a fluid.

THE STABILITY OF A FLUID depends on the vertical distribution of density within it. A fluid is said to be *stable* if, when a sample or parcel of it is given a small vertical displacement, there is a restoring force which acts on the parcel to return it to its original level. Conversely, it is said to be *unstable* if the parcel, once displaced, experiences a force moving it further away from its original level, and a state of *neutral stability* is said to exist if it experiences no such force in either direction at its new level. These three states can be illustrated by considering what will happen to a ball displaced on a concave, a convex, and a horizontal surface as illustrated in Figure 6. 1.

In a fluid there are three principal ways in which density may vary with depth (or height) (Fig. 6. 2). In Case *A*, where density increases with depth, a parcel of fluid which is displaced vertically up will, *if its density remains constant during the displacement*, be denser than the surrounding fluid at its new level and will thus tend to sink back to its original level. Had it been displaced downwards it would have been less dense than the surrounding fluid and thus it would have again tended to return to its original level. This is the situation which we have defined as stable. Similar considerations show that, if the density of the fluid remains constant during a vertical displacement, Case *B* is an unstable situation, and Case *C* is one of neutral stability. However, when a fluid is displaced vertically it will experience a different pressure, which will itself bring about a change in density due to the compressibility of the fluid, and will also bring about an adiabatic temperature change which further affects the density of the fluid.

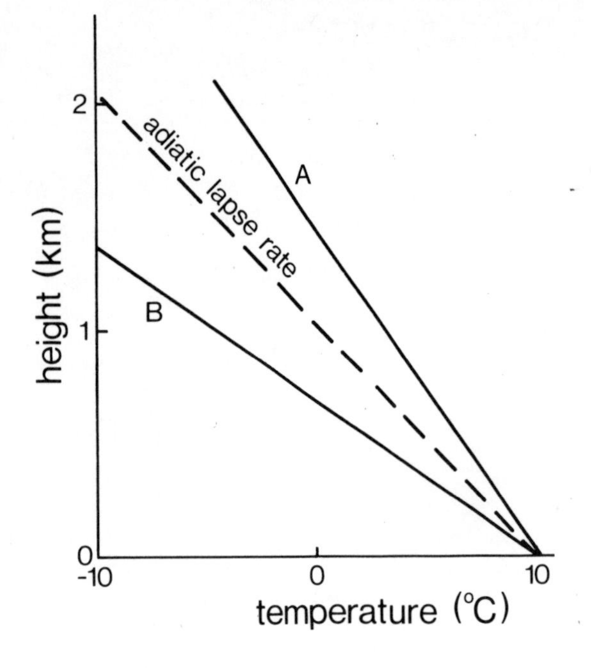

6.3 Stable (A) and unstable (B) environment curves in the atmosphere.

STABILITY IN THE OCEANS

The compressibility of sea water is a function of both temperature and salinity, but we will ignore these effects in our treatment here. The direct effect of pressure on density can be eliminated if we consider the density which the water would have if it were at the sea surface (i.e. at atmospheric pressure) with its observed temperature and salinity. Oceanographers use the quantity σ instead of density as a form of shorthand. It is related to density measured in gm cm^{-3} (ρ), or to specific gravity which is numerically the same, by the expression:

$$\sigma = (\rho - 1)\,10^3 \,. \qquad (6.\mathrm{I})$$

Thus the σ value equivalent to a specific gravity of 1.02783 would be 27.83. The σ value which relates to the specific gravity at atmospheric pressure is denoted by σ_t. The variation of σ_t with depth z ($d\sigma_t/dz$) thus provides a first approximation of the stability of a water column in the ocean. A closer approximation which takes account of adiabatic temperature changes is obtained by using $d\sigma_\theta/dz$, where σ_θ is the σ value relating to specific gravity at atmospheric pressure and at the observed salinity, but at the potential temperature θ, i.e. the temperature which a parcel of the fluid would have if it were brought adiabatically from its initial state to the standard atmospheric pressure of 1 000 mb.

STABILITY IN THE ATMOSPHERE

The density of air depends essentially on just two factors, pressure and temperature. An increased amount of water vapour in air does reduce its density, but for most practical purposes this can be ignored. At a particular pressure level in the atmosphere, then, density is a function only of temperature, and we can thus assess the stability of a column of air in the atmosphere by plotting its temperature against pressure. Such a plot is known as an *environment curve*. To decide on the stability of the air column, however, we must take account of the adiabatic change of temperature which will take place when a parcel of air is displaced vertically, and we must therefore compare this environment curve with the adiabatic lapse rate. In Chapter 2 we stated that the adiabatic lapse rate for air which is not saturated with water vapour is 9.8°C km^{-1}, and as this is stated in relation to height, it will be more convenient here to consider the environment curve plotted as temperature against height rather than pressure. Figure 6. 3 shows examples of stable and unstable environment curves. When the environmental lapse rate is less than the adiabatic lapse rate (Case *A*), air which is displaced vertically upwards will cool at the adiabatic lapse rate so that it is cooler than the surrounding air, and thus will sink back to its original level; i.e. this is a stable situation. In Case *B*, where the environmental lapse rate is greater than the adiabatic lapse rate, air displaced vertically upwards will be cooled down adiabatically but will still be warmer than its surroundings and will thus continue to rise; i.e. the situation is unstable. It is left to the reader to work out what happens if air is displaced downwards in each case.

If the air is saturated with water vapour, or if it becomes saturated as it rises due to adiabatic cooling, further ascent and associated adiabatic cooling will lead to condensation. When this takes place the latent heat of vaporisation is released, and this partially offsets the adiabatic cooling. The reduced rate of cooling which results is the *Saturated Adiabatic Lapse Rate* (SALR), as distinct from the

Dry Adiabatic Lapse Rate (DALR) of 9.8°C km^{-1}. The SALR varies with temperature because a small temperature decrease at high temperatures produces much more condensation than the same temperature decrease at low temperatures, e.g. at 20°C and 1 000 mb the SALR is 4.3°C km^{-1} whilst at −20°C and 1 000 mb it is 8.6°C km^{-1}. A full consideration of atmospheric stability must therefore compare the environment curve with both the DALR and the SALR (Fig. 6. 4), and for the air to be stable regardless of its water vapour content the environmental lapse rate must be less than the SALR. If the environmental lapse rate is between the DALR and the SALR the air is said to be *conditionally unstable* – conditional upon it being saturated. Air which is conditionally unstable with a relative humidity of less than 100% would therefore behave as stable air, but it could become unstable if it was forced to rise, e.g. by an orographic (hill or mountain) barrier, so that it reached its condensation level. Meteorologists plot observations from sounding balloons of temperature and humidity as a function of pressure on adiabatic diagrams, e.g. the tephigram (Fig. 6. 5), to identify the level at which a lifting mass of air will condense or the level to which mixing must take place for condensation to occur, and to assess the stability of the air. From this the likelihood of cloud development, its type, and the height of its base and top can be predicted. In general, stratus-type cloud develops in stable air which is forced to rise, and cumulus-type cloud develops in unstable air. Lee waves (Fig. 3. 4) form in stable air which passes over an orographic barrier, and similar features known as internal waves develop in stable layers in the ocean.

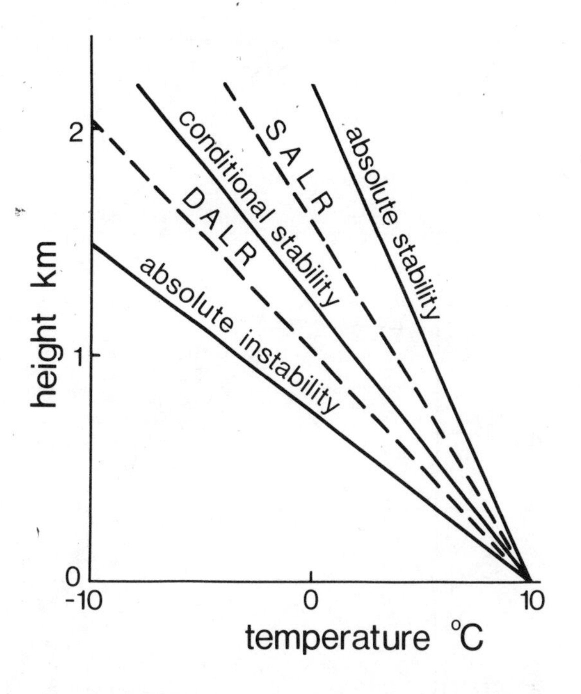

6. 4 Stability of environment curves in the atmosphere in relation to DALR and SALR.

FÖHN WINDS

Warm winds which are experienced in the lee of high ground often owe their anomalously high temperatures in part to the difference between the DALR and the SALR. Moist air rising over the high ground quickly reaches its condensation level and then cools more slowly at the SALR. If precipitation falls from the clouds which develop on the windward slopes, there will be relatively little water present in any clouds on the lee side, and this will evaporate quite quickly. The air will then warm at the DALR as it descends (Fig. 6. 6). Such winds, which are very dry as well as warm, occur on the northern side of the Alps where they are known as *Föhn*, and on the eastern side of the Rockies where they are called *chinook* winds. But even when no condensation and precipitation have occurred on the windward slopes, the air on the leeward side is often warmer than that on the windward side, and this usually appears to be attributable to vertical interchange within the air as it passes over the high land.

ENVIRONMENTAL LAPSE RATE IN THE TROPOSPHERE

In the lower part of the atmosphere, air temperature normally decreases with increasing height. However, above a level which varies from about 9 km over the poles to 17 km over the equator, and which also varies seasonally and over shorter periods, air temperatures become constant or increase with increasing height. This level is known as the *tropopause*, and the layer beneath it as the *troposphere*. The average environmental lapse rate in this layer is about 6.5°C km^{-1} – appreciably less than the DALR, but comparable to the SALR. This decrease of temperature results in part from the adiabatic cooling of rising air, but also from the increasing distance from the

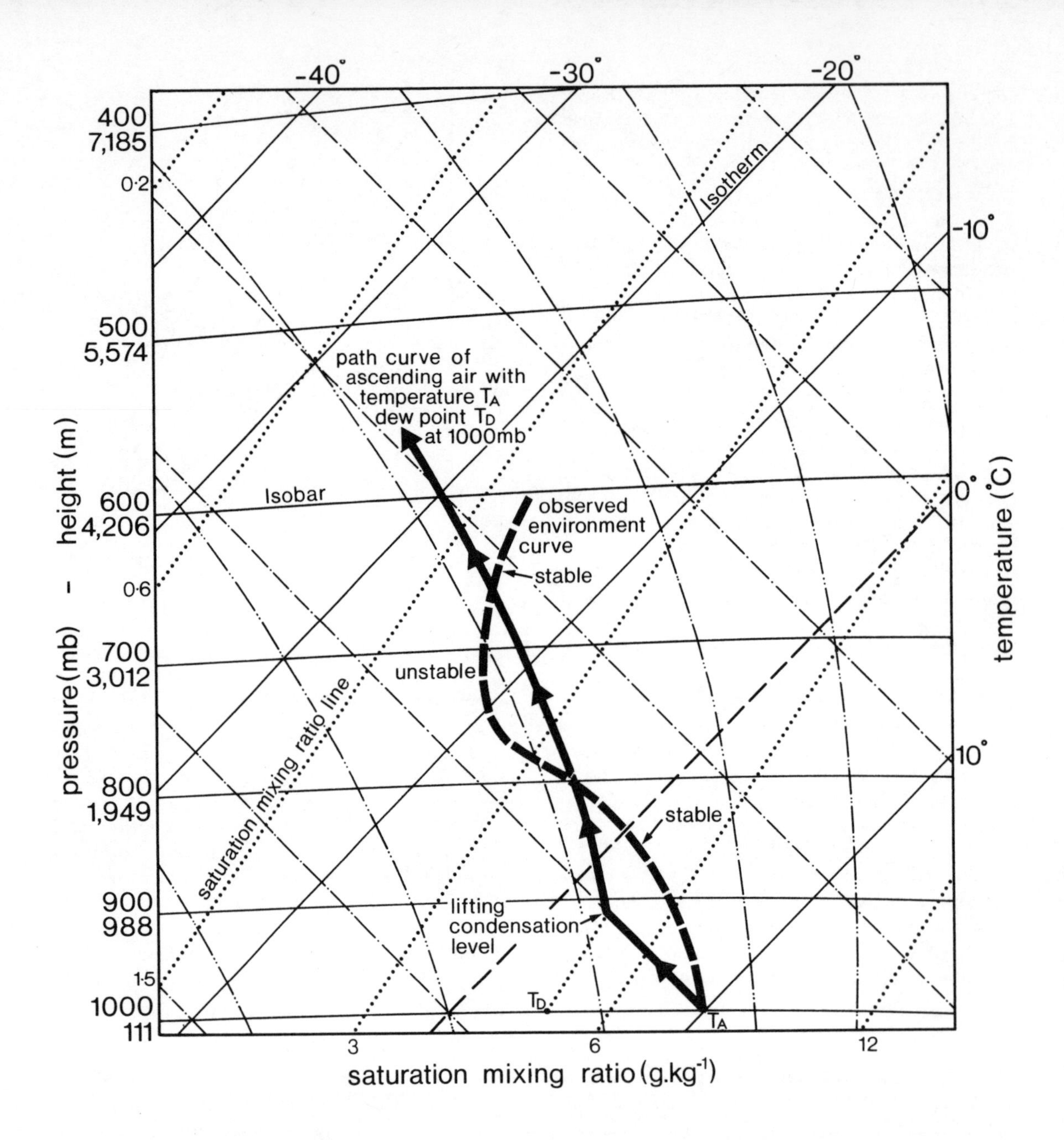

6. 5 A 'tephigram' illustrating an observed environment curve and the path curve of air ascending from the surface, and indicating the layers in which the air will be stable and unstable.

Earth's surface and the decreasing concentration of water vapour with height – the Earth's surface is the effective heat source for the lower atmosphere, and water vapour is the most effective atmospheric constituent for absorbing radiation.

The decrease in air temperature up a mountainside, which may be referred to as the *topographic lapse rate*, can differ greatly from the environmental lapse rate in the free air. The effective heat source has now been raised, but the atmos-

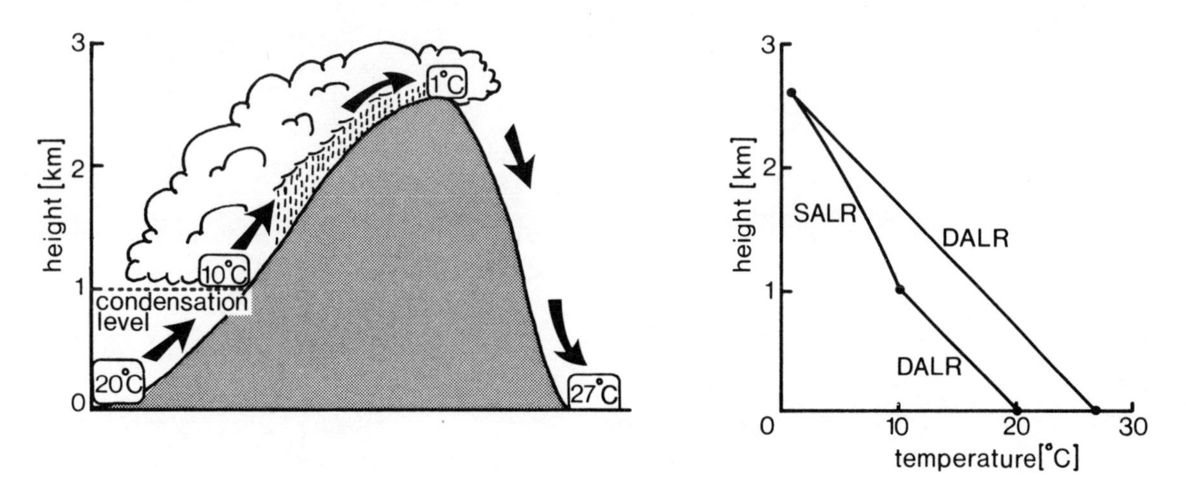

6. 6 The Föhn effect when air is forced over a mountain barrier and precipitation occurs on the windward slopes.

phere above it is less dense and thus less radiation is absorbed by the air. During a sunny day, solar radiation will be most intense on the highest parts of mountains, and so air temperature close to the ground may increase with height, but at night, when terrestrial radiation escapes from the summits with little impedance from the atmosphere, temperatures can fall to very low values, giving a large diurnal range at high altitudes.

There can be considerable deviations from this average environmental lapse rate. Sometimes air temperatures in the troposphere increase with increasing height. Such a *temperature inversion* gives very stable conditions. It may occur immediately above the surface (Fig. 6. 7-a), because of the cooling of the lowest layers of air by the ground surface – the conditions which cause fog. Alternatively it may occur in a layer well above the surface (Fig. 6. 7-b) caused either by the advection of warm air over a layer of cold air, or by air subsiding and warming at the DALR and then spreading out over cool air below. Such a stable layer in the atmosphere provides a ceiling for convection which is generated below it and thus limits the height of any cloud development, so that any clouds which do form must be of the stratus-type. Lapse rates exceeding the DALR can result from intense heating of the ground surface by solar radiation. Convection takes place in such unstable conditions, which are thus associated with cumulus-type clouds, showers, and perhaps thunderstorms.

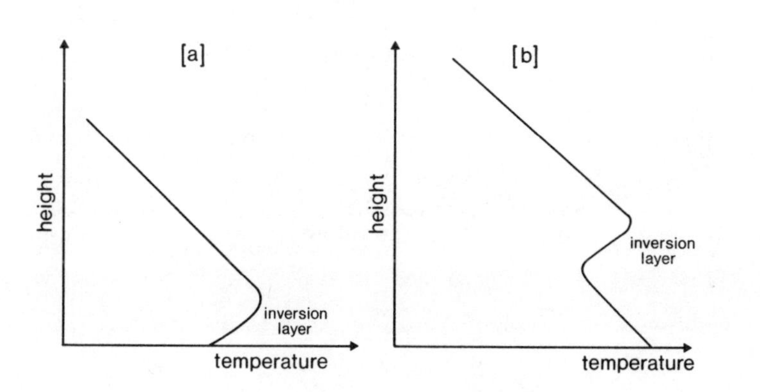

6. 7 Examples of temperature inversions in the atmosphere.

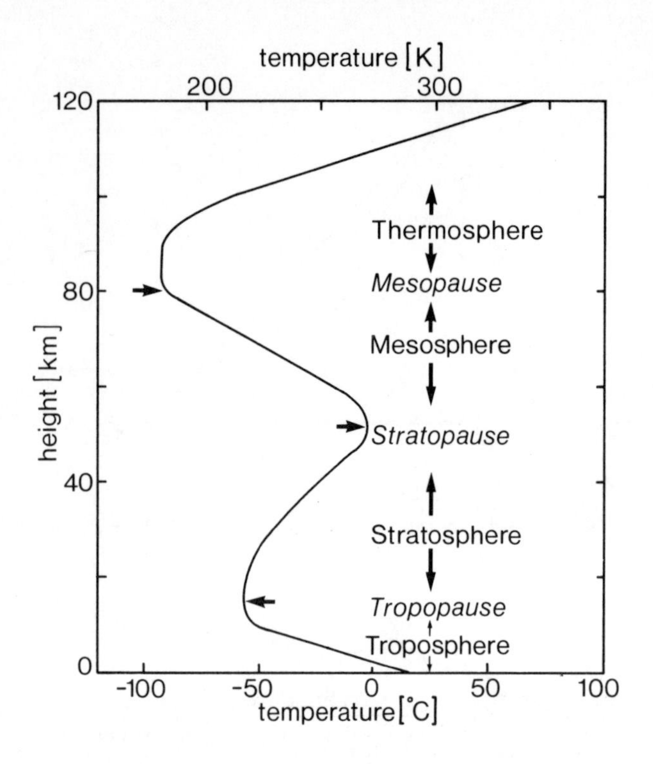

6.8 The typical variation of temperature with height in the atmosphere, and the names of the thermal layers (-spheres) and levels separating them (-pauses).

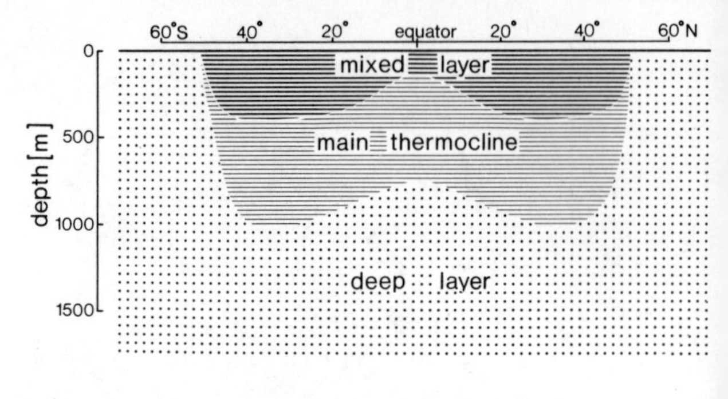

6.9 The thermal layers of the ocean.

TEMPERATURE STRUCTURE OF THE UPPER ATMOSPHERE

The tropopause marks the base of a major inversion layer in the atmosphere, the stratosphere. It therefore limits the height of cloud growth and causes cumulonimbus clouds which reach it to spread out laterally, forming their characteristic anvil head (Fig. 3.6). The temperature at the tropopause is of the order of −60°C, and the higher temperatures above in the stratosphere are attributable to the absorption of solar radiation there by ozone (Chap. 5), and in some situations to subsidence at the DALR. The stratosphere extends up to the stratopause at about 50 km, where the temperature reaches about 0°C. This is well above the level of maximum ozone concentration, but at this level almost

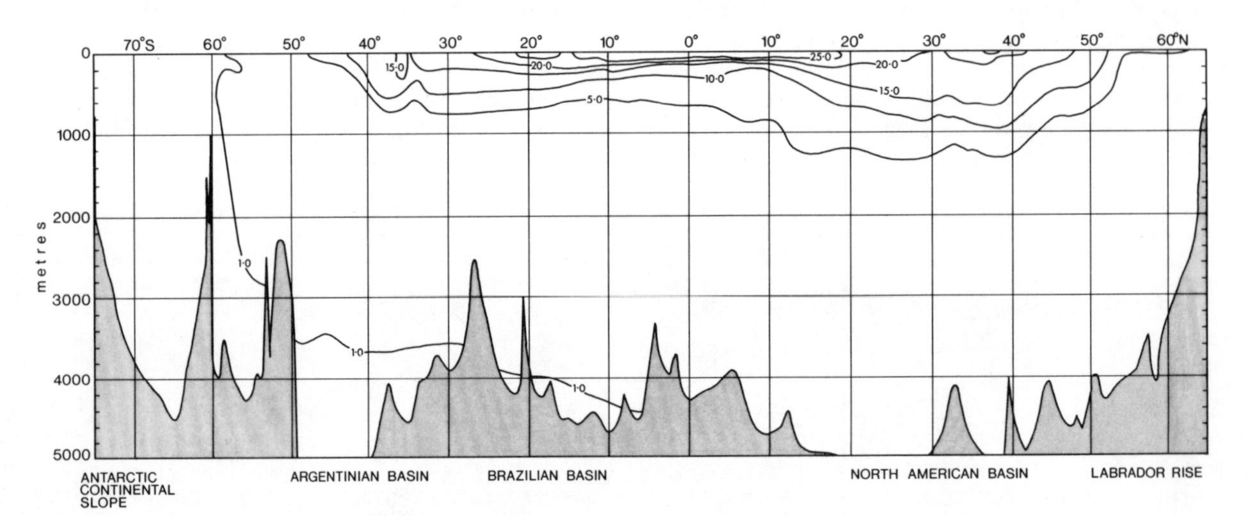

6.10 A meridional temperature section through the Western Trough of the Atlantic Ocean. (After G. Wuest)

all of the ultra-violet rays are still present in the solar radiation and thus considerable absorption takes place.

Temperatures decrease once again with height in the *mesosphere*, which extends from the stratopause to the mesopause at about 80 km. Above this is the *thermosphere*, where the very shortest wavelengths of solar radiation are absorbed. This includes the ionosphere which was discussed in Chapter 5. As in the stratosphere, temperatures here increase with height as one gets nearer to the source of the radiation, but as the density of the air at these levels is extremely low, its heat capacity is minute, and thus the temperatures have little practical significance.

Figure 6. 8 summarises the main features of the vertical thermal structure of the atmosphere to a height of 120 km, but it should be remembered that there are considerable departures from the mean according to the latitude, the season, and the time of day.

THERMOCLINES AND THE TEMPERATURE STRUCTURE OF THE OCEANS

The oceans are heated in their surface layer, which leads to a characteristic decrease of temperature with depth – a stable situation. This decrease is not usually gradual, but takes the form of a major step, or sometimes a series of steps, in which the thermal gradient often exceeds 5°C $(100\ m)^{-1}$. Such steps, which are layers of considerable stability, are named *thermoclines*. When the thermal gradient is small, the water column is only just stable and is readily susceptible to turbulent mixing, which may be generated by the wind in the surface layers or by shear in ocean currents below. A layer which is mixed in this way becomes isothermal (apart from the small adiabatic temperature gradient), but temperature differences develop between this layer and those above and below which have not been involved in the mixing. Thermoclines which are formed in this way are then able to resist mixing due to their stability, and the downward transfer of heat through them can only be accomplished by very slow processes of diffusion.

Three types of thermoclines are recognised in the ocean. The *main thermocline*, which is a permanent feature in middle and low latitudes, and *diurnal* and *seasonal thermoclines* which are destroyed by cooling at the ocean surface. The main thermocline is associated with the general circulation of the oceans which will be discussed in Chapter 14. The seasonal thermocline is essentially a feature of mid-latitudes, being formed in the spring and summer, usually at depths of between 50 and 100 m. The net cooling at the surface in autumn and winter, together with strong winds, lowers the thermocline but reduces the temperature step across it until the water is fully mixed down to the main thermocline at perhaps 300 m. In low latitudes there is insufficient seasonal contrast to produce this phenomenon, and the main thermocline is accordingly nearer to the surface at depths of perhaps 100 or 150 m. In higher latitudes, the main thermocline is pushed deeper by the more intense winter cooling until it disappears completely at latitudes greater than about 60°. Seasonal thermoclines still develop, however, particularly where there is a relatively fresh layer of surface water providing an initial step in the density gradient. Diurnal thermoclines may develop in any latitude when there is a sufficient input of heat during the day. Their temperature step may be 1 or 2°C, and they are typically found at depths of about 10 m, but this depends essentially on the wind conditions.

The main thermocline allows the ocean to be divided into three principal layers (Fig. 6. 9), and if winter conditions are considered to exclude diurnal and seasonal thermoclines, this pattern can readily be recognised in meridional temperature sections through the oceans (Fig. 6. 10). In the main thermocline, temperatures fall from more than 15°C to about 5°C, and in the deep water they decrease slowly from 5°C to less than 2°C. Names have been proposed for these layers of the ocean following the terminology used in the atmosphere – oceanic troposphere for the surface mixed layer, tropospheric discontinuity for the main thermocline, and oceanic stratosphere for the deep water – but the analogy with the atmosphere is incomplete and they have not come into general use.

It should finally be noted that, in fresh water, colder water can overlie water at 4°C in a stable situation, and thus an 'inverse thermocline' can develop in the winter months in a lake which cools to less than 4°C, and a temperature profile in a lake will not show temperature decreasing with depth below 4°C. In a stratified lake the surface mixed layer is known as the *epilimnion*, and the deep layer below the thermocline as the *hypolimnion*.

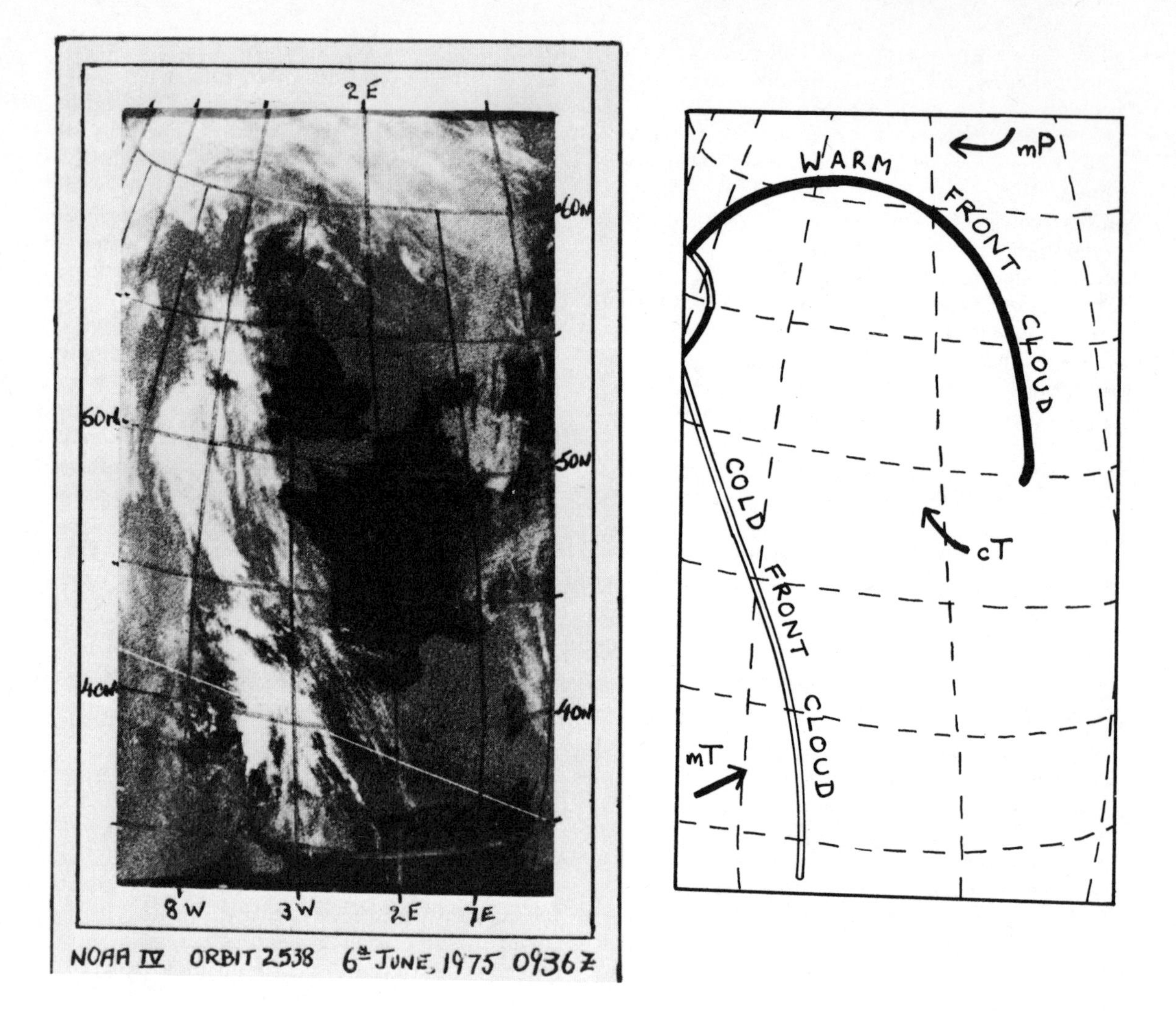

7.0 A NOAA IV satellite photograph of the weather situation above the British Isles, France and the Western Mediterranean, and the corresponding 'overlay' indicating the fronts and the mP (maritime polar), cT (continental tropical) and mT (maritime tropical) air masses.

7 Air Masses and Water Masses

The air mass concept ☐ Characteristics, source regions, modification ☐ Classification and associated weather with examples from British Isles ☐ The extension of this concept to water masses ☐ T-S diagrams ☐ Conservative and non-conservative properties ☐ Examples of water masses, and their rôle in tracing sub-surface water movement in the oceans.

THE AIR MASS CONCEPT

Weather observations began to be sufficiently plentiful in the second half of the last century for synoptic maps to be constructed showing the surface distribution of weather elements such as temperature, pressure, winds and precipitation. Analyses of these led to the introduction of the concept of *air masses*, which allowed the individual elements to be integrated, weather situations to be identified, and weather forecasts to be made. Upper air observations in more recent years have added considerably to our understanding, but the air mass concept has proved to be valid and continues to form the basis of most synoptic meteorology and climatology.

An air mass is now defined as a large body of air, with horizontal dimensions of at least hundreds and perhaps thousands of kilometres and a vertical dimension of perhaps 5 km, which has more or less uniform horizontal characteristics of temperature and humidity. It has fairly distinct boundaries separating it from other air masses. These are the *fronts*, and they usually slope at quite small angles to the horizontal so that one air mass will overlie another within the troposphere. A front is identified by a marked change in temperature, humidity and wind direction along a horizontal surface, either at ground level or above.

SOURCE REGIONS

An air mass acquires its distinctive characteristics in a source region where there is a large and fairly uniform surface, either water or land, over which air remains fairly stagnant for a period of at least a few days. Source regions are most commonly areas over which air is subsiding and then spreading out laterally, and fronts are situated in areas of convergence where air masses from different source regions meet. In moving away from their source regions, air masses will be modified by the surfaces over which they pass, and thus their later properties will depend not only on their source region but also on the nature of the surface they pass over and their age since being formed.

Air masses are classified primarily according to their source region, which is considered to come

Table 7. 1: Air mass classification and properties with examples of source regions

<table>
<tr><th></th><th>Tropical</th><th>Polar</th><th>Arctic/Antarctic</th></tr>
<tr><td>Maritime</td><td>maritime tropical (mT)
warm and very moist;
near Azores in
N. Atlantic</td><td>maritime polar (mP)
cool and fairly moist;
Atlantic south of
Greenland</td><td rowspan="2">arctic or antarctic
(A) (AA)
very cold and dry;
frozen Arctic Ocean
central Antarctica</td></tr>
<tr><td>Continental</td><td>continental tropical (cT)
hot and dry;
Sahara desert</td><td>continental polar (cP)
cold and dry;
Siberia in winter</td></tr>
</table>

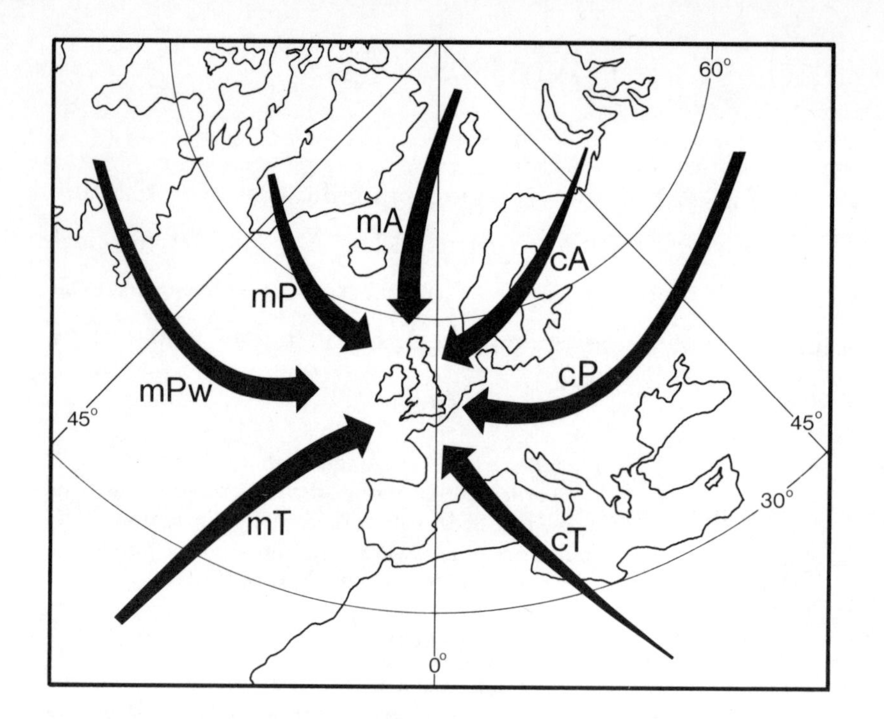

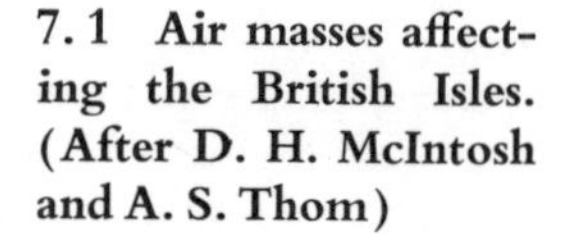

7.1 Air masses affecting the British Isles. (After D. H. McIntosh and A. S. Thom)

within one of three latitude belts (Arctic/Antarctic, Polar or Tropical), and to be either maritime or continental. Table 7.1 summarises the principal types which are thus defined, together with their properties at source. 'Polar' is used to identify air masses coming from latitudes of 50° to 70°, while 'Arctic' and 'Antarctic' are used for air masses originating near the poles.

Both Arctic and Antarctic air masses are of an essentially continental nature as they are formed over frozen surfaces, though they may subsequently be modified by passing over a stretch of unfrozen ocean. A maritime equatorial air mass (mE) is also recognised by some authors, but as the equatorial region is one of convergence the air masses which are found here have usually come from the sub-tropics.

AIR MASS MODIFICATION

The effects of a surface over which an air mass passes are felt in its lower layers. They may bring about changes in the moisture content of the air through evaporation or precipitation, and also in its temperature by the release of latent heat or by heat exchange with the surface. Changes aloft result from air movement, and are thus known as *dynamic* changes. The horizontal movement of the air will almost certainly vary with height so that the air mass will not move like a rigid entity, and this shear will induce turbulent mixing. Further, if the lower layers of the air mass are being warmed it will become increasingly unstable, and convective mixing will develop. Other dynamic changes are associated with large-scale vertical motion of the air.

The modifications which an air mass is experiencing can be indicated by adding a further letter after the air mass symbol. If the lower layers of an air mass are warmer than the surface over which it is passing the letter 'w' is added, but if they are cooler 'k' is added. Thus an mPw air mass is being cooled at the surface, increasing its stability, whilst an mPk air mass is being warmed and becoming unstable.

AIR MASSES AND ASSOCIATED WEATHER IN BRITAIN

Weather conditions in a location may be considered to stem from the particular air mass which is present and from the modifications which it is undergoing. Britain, being situated in a region of confluence in mid-latitudes, experiences the effects of most types of air mass (Fig. 7. 1) and thus provides a good example for the study of weather conditions produced by their surface modification. Dynamic changes, associated particularly with vertical air movements, are also very important in determining weather conditions, and must not be overlooked in any particular situation.

mP air reaching Britain is usually of the mPk type, and is thus becoming unstable. It will, however, have maintained a high relative humidity by evaporation during its passage over the ocean, and thus cumulus and cumulonimbus clouds with showers will develop, particularly over warm land in the afternoon. Temperatures will be below average, so that in winter the showers will often be of snow or sleet. The gusty winds and convection will disperse dust and smoke so that visibility will be good, giving bright intervals between showers.

If mP air follows a more southerly route from its source region and then recurves to approach Britain from south of west, it may well be of the mPw type, sometimes called 'returning polar maritime air'. It brings normal temperatures and weather intermediate between that of mPk and mT air.

mT air is invariably of the mTw type and is thus stable; having approached Britain across the ocean and been cooled, it is saturated or near-saturated with water vapour. It brings mild weather with dull skies and poor visibility, and not infrequent fog on the western side of Britain. Where it is forced to rise over orographic barriers, stratus cloud develops and drizzle or heavier rain is common, with a rain shadow to the east of the mountains.

cT air is unstable in its source region, and although its lower layers become stable on its passage to Britain, the upper layers remain unstable and thunderstorms can develop in summer. In winter, however, its lowest layers are very stable and any clouds which develop are usually stratus type. In general it brings temperatures well above average with hazy conditions.

cP air brings very cold conditions in winter. It is stable in its source region, but may become unstable in its lower layers and gain a considerable water vapour content in passing over the North Sea. Clouds are of the cumulus type, though they may be restricted to stratocumulus, and in winter showers of sleet or snow may be experienced in the eastern half of Britain.

Arctic air may be considered as cA or mA according to the route which it has followed from its source region to Britain. cA air reaches Britain after passing over Scandinavia. It is similar to cP air, though it is even colder and is more likely to bring snow in winter and spring. mA air follows a route over the Greenland and Norwegian seas and is comparable to mPk air though colder and even more unstable. Heavy snow showers, prolonged frost and exceptionally good visibility are all characteristic of it in winter and spring.

WATER MASSES AND THE T–S DIAGRAM

Oceanographers have made use of a concept similar to that of air masses, and discern water masses essentially by their temperature and salinity characteristics. They, too, are considered to form in a source region where they are in the surface mixed layer, and where they are influenced by a fairly constant set of atmospheric conditions. If the water is considered to remain stationary for a sufficiently long period of time, the difference between evaporation and precipitation, together with the input of fresh water from rivers in coastal regions and the results of ice melting and sea ice forming in high latitudes, will determine the salinity of the water. Similarly its temperature will result from the radiation balance at the water surface together with heat exchange with the atmosphere. If the salinity of the water is being decreased and its temperature increased, the density of the water will be lowered, the water column will be stable, and only a very shallow water mass can be formed. If, however, the salinity is increased and the temperature decreased, the water becomes denser and sinks, so that a water mass of considerable vertical extent may be formed.

In order to identify water masses, a set of observations of temperature and salinity at successive depths at one position in the ocean is plotted on a graph with temperature as the ordinate and salinity as the abscissa, and the points are joined up in order of increasing depth (Fig. 7. 2). If a water mass is completely homogeneous it will be represented by a single point on this diagram, and in this case it is described as a *water type*. The clustering of observations about this point will indicate the presence of this water type. But water masses often show a variation of temperature and salinity with depth and are characterised by a particular curve on the T-S diagram. This variation may result from waters of slightly different properties forming at differing times of year and sinking to depths appropriate to their densities. Alternatively, conditions may vary at the surface within the source region and the water may sink not vertically but along sloping surfaces of equal density (*isopycnals*). As σ_t is a function of temperature and salinity only, lines of constant σ_t can be shown on the T-S diagram as in Figure 7. 2. A first indication of the stability of a water column can then be obtained by comparing the T-S plot with the trend of the σ_t lines.

CONSERVATIVE AND NON-CONSERVATIVE PROPERTIES

Once formed, water masses like air masses move away from their source regions and are subject to modification. If they remain in, or return to, the surface mixed layer, further interaction with the atmosphere will bring about changes in their temperature and salinity. Alternatively mixing with other adjacent water masses will produce water with properties intermediate to those of the parent masses. Once a water mass is no longer subject to interaction with the atmosphere, its temperature and salinity can only be changed by mixing processes, and these properties are therefore said to be *conservative*.

Water masses usually have particular chemical characteristics and organisms associated with them, as well as a typical T-S relationship. Dissolved oxygen concentration is often a useful indicator identifying a water mass, and so are the concentrations of certain nutrient salts such as silicates and phosphates. The organisms associated with particular water masses are known as *indicator species*. They may remain with the mass because its physical and chemical properties suit them, or simply because they are free-floating and are carried with it when it moves away from its source region. These properties, however, are subject to change by chemical and biological processes in the ocean, and hence they are referred to as *non-conservative properties*.

EXAMPLES OF WATER MASSES

Water masses which form in semi-enclosed basins provide particularly clear examples of the concept. That which forms in the Baltic Sea is of low salinity due to a considerable excess of precipitation and river run-off over evaporation. In summer it is also warm and thus of very low density. It flows out of its source region through the restricted sounds between Sweden and Denmark where intense mixing takes place with inflowing water below. Before mixing, its temperature in summer is about 16°C and its salinity less than 8‰, but by the time it reaches the Skagerrak its salinity has been increased by mixing to some 20‰. Remaining at the surface due to its low density, it is rapidly modified by further interaction with the atmosphere, and thus its influence does not reach to any great distance beyond the Baltic.

In the Mediterranean Sea, evaporation exceeds the input of fresh water by precipitation and run-off, and so salinity increases. In the north-western Mediterranean in winter, cooling (associated particularly with the wind called the Mistral) can lead to convection right to the sea floor at more than 2 000 m depth, so forming a deep homogeneous water type of salinity more than 38.4‰ and temperature about 12.8°C. In leaving the Mediterranean through the Straits of Gibraltar, this water is also subject to intensive mixing, and the least mixed layer or *core* of the Mediterranean water in the adjacent Atlantic has a salinity of 36.5‰ and temperature 11°C. Due to its high density it sinks to a depth of about 1 000 m. It spreads out at this level, being continuously modified by mixing, but its core is recognisable throughout much of the Atlantic Ocean.

In the open ocean, Central Water Masses form in about latitudes 25° to 40° and sink down along sloping isopycnals to occupy the upper part of the main thermocline. That in the North Atlantic is represented by a T-S curve which runs from 19°C and 36.7‰ to 8°C and 35.1‰. In somewhat higher latitudes Intermediate Water Masses form, which are characterised by low salinities but also by lower temperatures. The Antarctic Intermediate Water is the most widespread of these. It has temperatures between 2° and 7°C and salinities between 34.1 and 34.6‰, and spreads northwards after sinking at about 50°S to depths of 800 to 1 000 m. The deepest water masses are formed in the highest latitudes where water is cooled down to very low temperatures in winter, often to freezing point, so that the salinity is maintained by the freezing process (Chap. 4). Antarctic Bottom Water is essentially a water type with a temperature of −0.4°C and salinity 34.66‰, and spreads northwards in the oceans

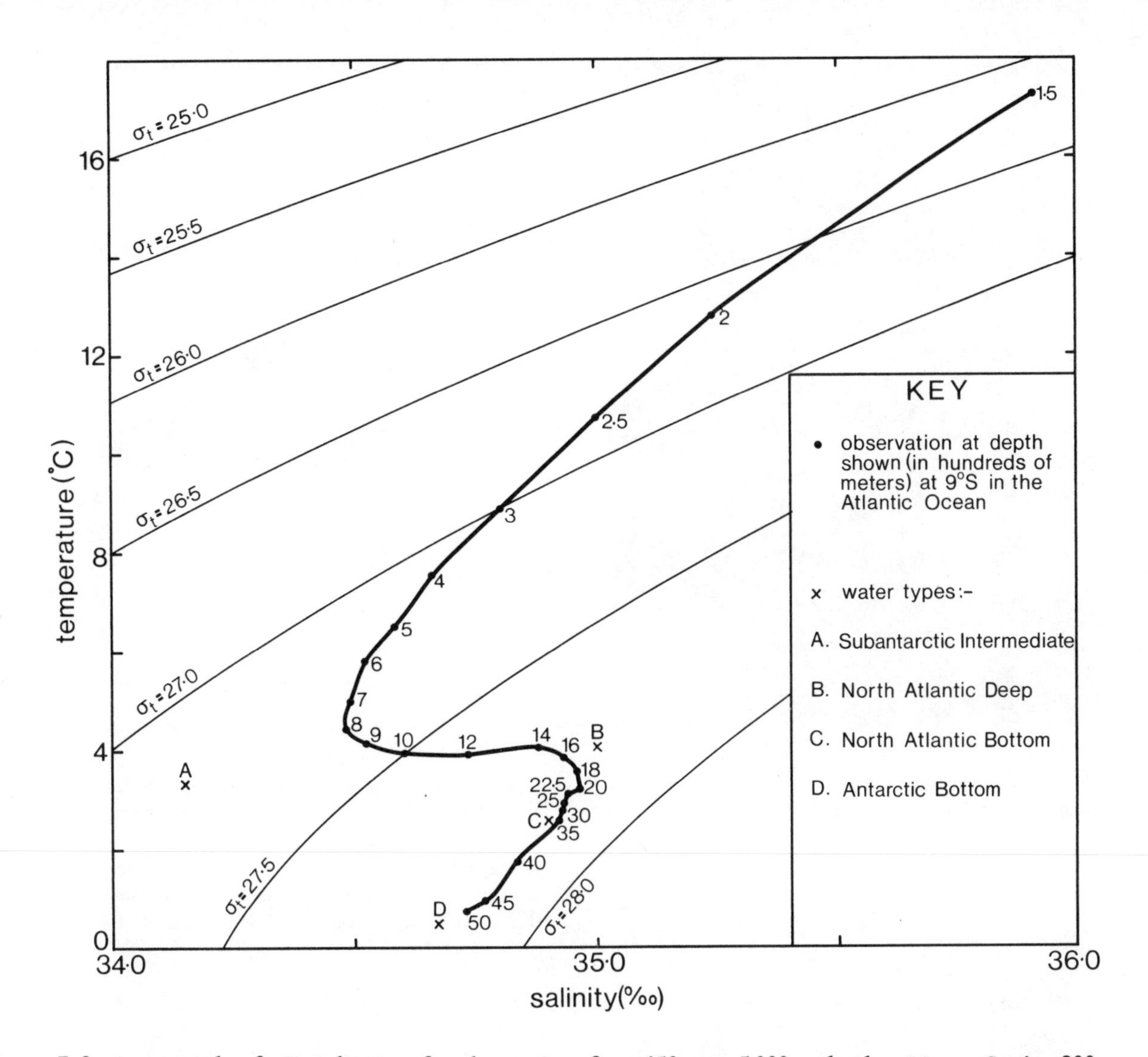

7.2 An example of a T-S diagram for observations from 150 m to 5 000 m depth at Meteor Station 200.

below 3 000 m. North Atlantic Deep and Bottom Waters, which have originated in the Norwegian and Greenland Seas but have been appreciably modified by mixing in flowing over the Scotland-Greenland Ridge, spread south and overlie Antarctic Bottom Water in the Equatorial and South Atlantic (Fig. 7.2).

The great value of the water mass concept has been in building up a descriptive picture of the circulation pattern of the oceans. The currents in the deep parts of the ocean are either too slow or too variable for the pattern of the average conditions to be obtained by direct observations, but water mass analysis allows the depth of the core of a water mass and the direction in which it is spreading to be identified. However, to deduce the *rate* at which it is moving, other data such as the rate of mixing and the rate of change of a non-conservative property are needed, but they are not usually available.

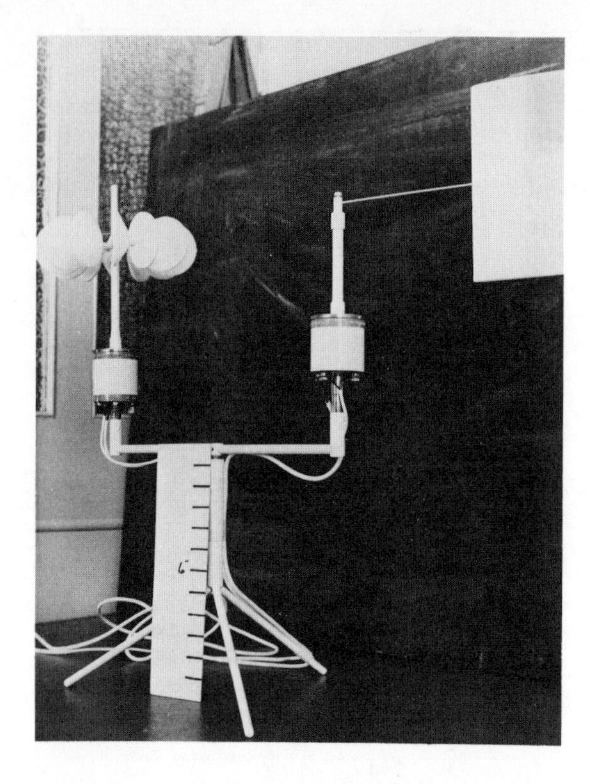

8. 0-a A sensitive photoelectric anemometer and wind-vane.

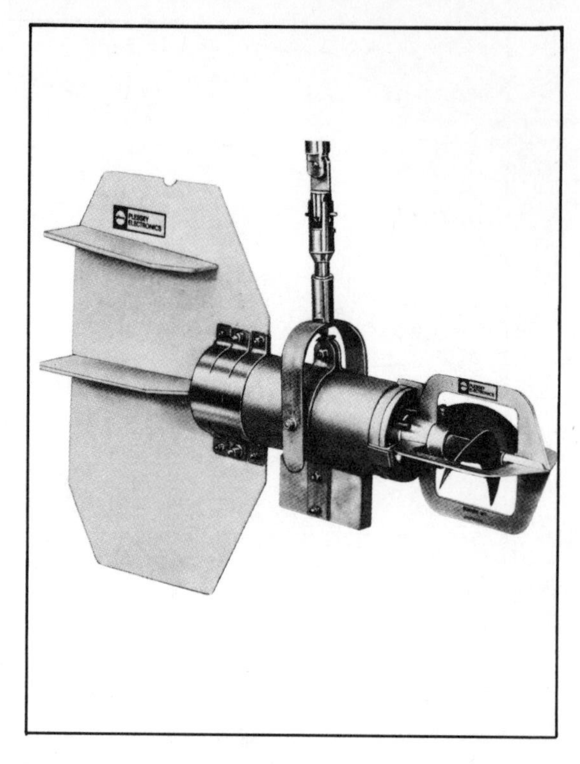

8. 0-b An oceanographic current meter which records current speed and direction internally on magnetic tape. It can stay submerged for up to 80 days (depending on the frequency of sampling), and can be connected to a deck print-out unit via a cable or an acoustic link.

8 Motions and Forces

Relationship between force and motion □ Steady state conditions □ Laminar and turbulent flow □ Scales of motion □ Types of forces involved □ Centrifugal force □ Coriolis Force.

ACCORDING TO NEWTON'S LAWS OF MOTION a body continues in a state of uniform motion (which includes the case when it is at rest) unless it is acted upon by a force or forces, when the rate of change of its momentum mv is proportional to the net force acting on it. This is expressed by the equation $F \propto d(mv)/dt$, or $F = ma$ (force $=$ mass$\times$acceleration, where acceleration may be positive or negative, i.e. a retardation) when appropriate units are employed. Motion is expressed in terms of velocity, which is a vector quantity having both magnitude (speed) and direction. (Note that, by convention, the direction of winds in the atmosphere is always expressed as the direction *from* which the wind blows, whereas the direction of currents in the ocean is always expressed as the direction *towards* which they flow. Hence, for example, a northerly wind blows in the opposite direction to that in which a northerly current flows.) A force acting upon a body may bring about a change in its speed, or in its direction, or both. Force is similarly a vector quantity, and when more than one force is acting upon a body the forces must be added vectorially (Fig. 8. 1-a) to obtain the net or *resultant* force.

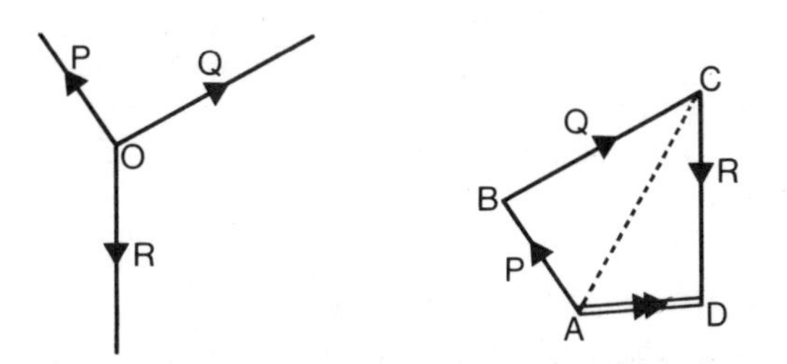

8. 1-a The vector addition of forces P, Q, R on the left gives the resultant force represented by the line AD on the right. If only P and Q were to be added, the resultant would be the vector AC.

So motions in the atmosphere and ocean result from the forces which act on the air or water. Vertical motion is dominated by the Earth's gravitational force and is related to the stability of the fluid column as discussed in Chapter 6. In the case of horizontal motion, there are usually a number of forces of comparable magnitude acting in concert. If the forces when added together vectorially balance and equal zero, a state of uniform motion will persist, but if there is a net force acting on the fluid it will be subject to acceleration. True uniform motion, however, would mean a constant velocity in relation to some co-ordinate system fixed in space. A particle on the surface of the Earth might appear to be in uniform motion relative to that surface, but in fact its motion is continually changing due to the rotation of the globe, and seems to be uniform only when observed relative to a system of co-ordinates which rotates with the Earth, e.g. a system based on latitude and longitude. If the framework of reference were fixed relative to, say, the centre of the Earth and to the directions in which we see distant stars, the motion of the particle would appear to be *not* uniform, and therefore to be subject to the continuous action of a force. To reconcile these seemingly conflicting observations we introduce the *Coriolis Effect*.

Consider a projectile fired in a direction due north from a gun on the equator (Fig. 8. 1-b). In addition to its muzzle velocity it will retain an eastward velocity which it shares with the gun and all things in the same latitude, because all these things are rotating to the east with the Earth's surface. In the absence of external forces the projectile must keep all its eastward velocity, but as it moves north the rotational speed of the ground under it lessens, and the projectile will draw ahead of the ground in its eastward motion. By the time it falls it may have deviated to the east

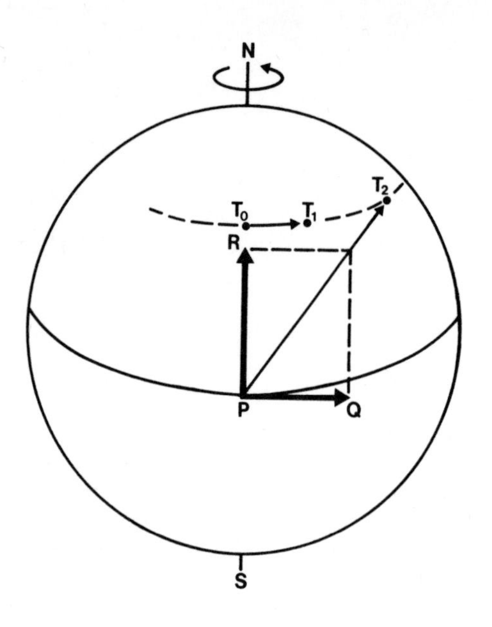

8.1-b A projectile is fired from P towards T_0 with velocity PR. During its flight it retains the eastward velocity PQ of its launch point and so moves towards T_2. Meanwhile T_0 moves eastwards more slowly to T_1. The displacement T_1T_2 is ascribed to the Coriolis Effect.

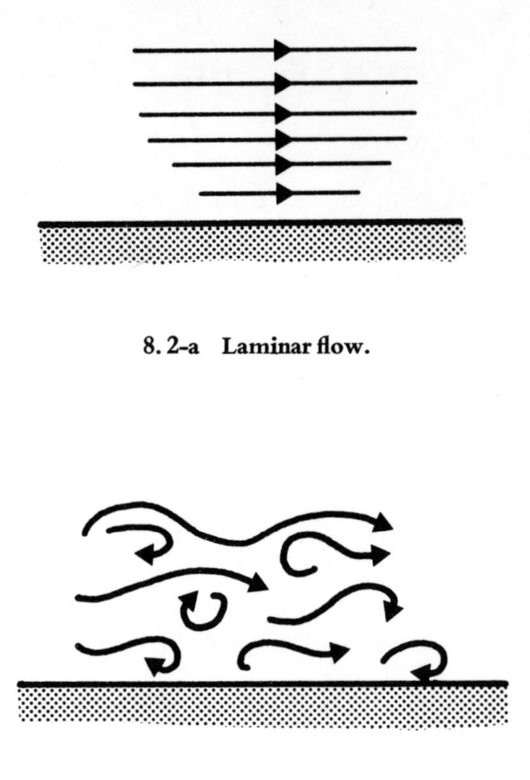

8.2-a Laminar flow.

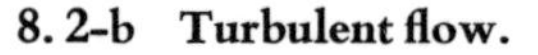

8.2-b Turbulent flow.

by several miles, *as if* a force had acted on it sideways. This apparent force is called the Coriolis Effect. If the projectile is fired *towards* the equator it will fly over ground which is moving progressively *faster* to the east, and the projectile will fall to the west of the line along which it was fired. This effect accounts for the apparent deflection of a moving particle, to the right in the northern hemisphere and to the left in the southern hemisphere; the effect is zero at the equator and maximal at the poles. Since its action is equivalent to that of a force, it is generally called the *Coriolis Force*. With this term taken into account, horizontal motion can be considered relative to the co-ordinate system which is provided by lines of latitude and longitude as though these were external rectangular co-ordinates fixed in space. The Coriolis Force affects the observed motion of *any* body moving relative to the Earth, and is not limited to motion with a north-south component in the horizontal plane. We return to this subject later in the chapter.

In this simplified treatment we shall consider mainly examples of motion where the forces acting upon air and water balance if the Coriolis Force is included, so that the velocity remains uniform relative to the rotating Earth. In reality the forces involved are continuously changing, and there are imbalances between them causing the air or water to experience accelerations relative to the Earth. In many cases, however, the average velocity remains nearly constant if suitable time periods are chosen for determining the averages, and this approximation is then found to give a reasonable treatment of the real situation.

LAMINAR AND TURBULENT FLOW

The types of motion which occur in the atmosphere and ocean may be classified in various ways. One of these is the distinction between *laminar* flow and *turbulent* flow (Fig. 8.2). In laminar flow the fluid moves in continuous sheets following parallel streamlines, whereas turbulent flow is disordered, with the streamlines of individual fluid elements criss-crossing one another. In a fluid of constant density the flow changes from being laminar to turbulent when the speed reaches some critical value which is directly proportional to the viscosity of the fluid and inversely proportional to its density and the distance from the nearest boundary of the fluid. In most situations in the oceans and atmosphere the flow is turbulent, and the effective viscosity or internal friction depends on the nature and degree of turbulence but is usually many orders of magnitude greater than the molecular viscosity. There are two types of situation, however, in which laminar flow may be experienced. One applies to the very shallow layer adjacent to the boundary of the fluid over a smooth surface, the other to regions of considerable vertical stability (such as atmospheric inversion layers and oceanic thermoclines) where the vertical components of turbulence are suppressed. The vertical current shears in such situations are very much greater than those in turbulent flows.

SCALES OF MOTION

An alternative method of classifying atmospheric and oceanic motions is based on the time and distance scales by which they may be recognised, and the extent to which their variations are periodic or irregular.

On the largest time and distance scales come the virtually continuous types of motion such as the Trade Winds or the Gulf Stream. Although these may show considerable fluctuations, they may be considered to be more or less permanent features, with distance scales of the order of thousands of kilometres.

Next come those motions which have a seasonal cycle. Most notable here are probably the monsoon winds of the Indian sub-continent and the associated seasonally reversing currents in the Indian Ocean. The distance scale of these is again of the order of thousands of kilometres, but their distinguishing feature is their periodicity.

Motions which extend over days or weeks are mainly irregular, and have distance scales of perhaps one thousand kilometres. They include the changes of wind which bring different air masses and weather to regions such as Britain, and similar and often associated changes in ocean currents.

When we come to motions with time scales from a few hours to one or two days, we find a wider variety, including some with obvious periodicity. This may be strictly diurnal, associated with the night-day variation in radiation balance as in land and sea breezes; it may be approximately semi-diurnal or diurnal as in tides; or it may vary somewhat as in travelling cyclones and other storms in the atmosphere. Distance scales for these types of motion range from under 50 km for land and sea breezes to 2 000 km for a mid-latitude depression.

On time scales of seconds, and less frequently minutes, come the regular features which are known as waves. The commonest are the wind-generated waves on the surface of the ocean with length scales of about 100 m, but longer waves exist in both the atmosphere and the ocean, e.g. the lee waves illustrated in Figure 3. 4. Irregular motions with periods of that order are associated with the turbulence in a flow, and show themselves for example in the gustiness of the wind.

The motion which is observed at a particular position in the ocean or atmosphere may be considered to be the vector sum of a number of velocities, each associated with one particular scale of motion, e.g. the velocity observed at one instant of time may be

$$\vec{u}_i = \vec{u}_1 + \vec{u}_2 + \vec{u}_3 + \vec{u}',$$

where $\vec{u}'$ denotes the component of velocity due to turbulence.

Often one can eliminate $\vec{u}'$ in the observation by means of a damping or averaging device in the instrument or method used. To find the velocity component related to one particular scale of motion requires that observations be made over a complete number of periods of all smaller scales of motion, so that the latter can be eliminated by vector averaging. This applies particularly to studies of ocean currents, where it is usually necessary to take readings over a complete number of tidal cycles so that the tidal component may be removed.

TYPES OF FORCES

The particular combination of forces which is the main cause of a particular type of motion may also be used to characterise that motion. This method, together with the scale of motion, will be used in the following chapters to identify forms of motion. It will be convenient to consider here

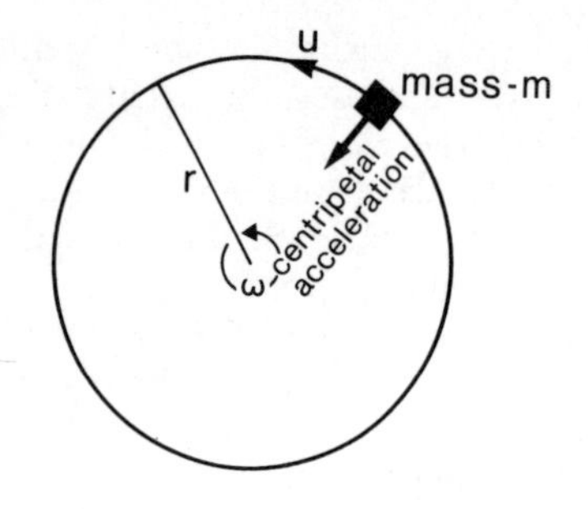

8. 3-a Centripetal acceleration.

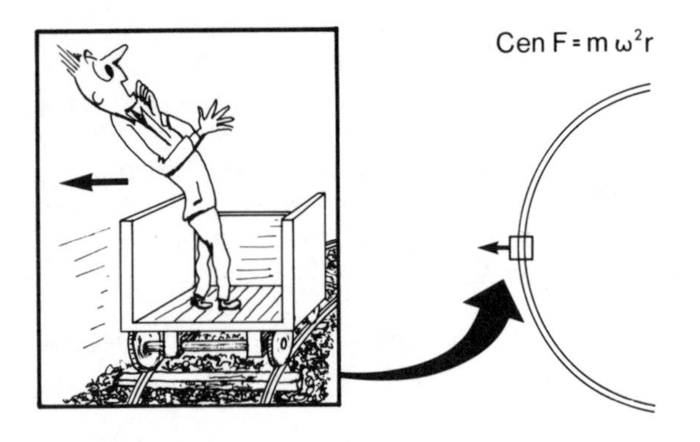

8. 3-b The apparent centrifugal force which an observer following a circular path experiences.

the various forces which may act in the atmosphere or ocean to cause or modify the horizontal movements of the fluid concerned.

Three categories of forces may be recognised: *external*, *internal* and *secondary*. The external forces arise from outside the fluid. The gravitational attractions of the Sun and Moon, which are responsible for tidal motions, come in this category, as does the force which moving air applies to the water in the oceans beneath. Internal forces result from the distribution of mass or density within the fluid. The uneven distribution of density stems essentially from the uneven heating of the oceans and atmosphere which was discussed in Chapter 5, and sets up horizontal pressure gradients within the fluid. By secondary forces are meant those which can only act upon the fluid when it is already in motion relative to the Earth's surface. Friction is the most obvious example, and it always acts in such a direction as to oppose the motion. Where different parts of a fluid are moving at different velocities, the friction between the parts, which is due to viscosity, acts so as to slow down the fastest moving parts of the fluid but to speed up the slower moving parts. When the flow is over a solid boundary, the friction will act in a direction opposite to the fluid's velocity immediately above the boundary. Although friction usually plays a minor rôle in atmospheric and oceanic motions, it would eventually bring them all to rest if internal or external forces did not continue to act, and so motion could not remain uniform in the absence of other forces. The other two secondary forces which require our attention are both *apparent* forces, resulting from the motion of the frame relative to which the velocity is being considered. They are the Coriolis Force (which has already been mentioned) and the centrifugal force which appears to act on a body following a circular path.

CENTRIFUGAL FORCE

A body travelling at a constant speed around a circle is changing the direction of its motion continuously, and is therefore undergoing an acceleration. This acceleration is directed towards the centre of the circle, perpendicular to the velocity vector at any instant of time, and is known as the *centripetal acceleration*. Hence for a body to continue on a circular path at a constant speed there must be a force acting towards the centre of the circle, and, as is shown in most elementary books on dynamics, the magnitude of this force is mu^2/r or $m\omega^2 r$, where m is the mass of the body, u its speed round the circle, r the radius of the circle, and ω the angular velocity of the body (usually measured in radians per second). (Fig. 8. 3-a). To an observer on the circling body, such as a passenger in a train on a circular track, the motion appears to be uniform, the ground going past at constant speed, except that he experiences an apparent force outwards from the centre of the circle, the *centrifugal force*. He balances this by leaning towards the inside of the circle, so setting up a *centripetal force*, which is in this case the horizontal component of the normal reaction exerted by the carriage seat or floor. In other words, for him to continue in this apparent state of uniform motion it is necessary for the centripetal force to be equal to this apparent centrifugal force in magnitude and opposite to it in direction (Fig. 8. 3-b). Centrifugal force will subsequently be referred to as CenF.

We shall confine ourselves here to the *horizontal* component of the Coriolis Force, or CorF for short, and consider it rather simply as the apparent force resulting from the Earth's rotation which must be balanced if a body is to move at a constant velocity relative to the Earth's surface. This enables us to determine its magnitude and direction fairly easily without introducing vector algebra. (A straightforward approach for those familiar with vector algebra will be found in many dynamics textbooks, e.g. Batchelor: *An Introduction to Fluid Dynamics*, Cambridge University Press, 1967.)

Consider a body, mass m, at position P on the surface of the Earth in latitude ϕ, moving with a speed u towards $\theta°$ east of north relative to the Earth (Fig. 8. 4). We may resolve the velocity of the body into an easterly component, u_x ($= u \sin\theta$), and a northerly component, u_y ($= u \cos\theta$), and treat each of these separately, finally combining their effects vectorially to obtain the total effect.

The easterly component, u_x: a body which is at rest *relative* to the Earth's surface is in fact rotating with the Earth and thus experiences a CenF outwards from the axis of rotation with a magnitude of $m\Omega^2 r$ where r is its distance from the axis of rotation and Ω is the corresponding angular velocity (Fig. 8. 5). This has a horizontal component towards the equator of magnitude $m\Omega^2 r \sin\phi$, and bodies on the Earth's surface would tend to move equatorwards were it not that the Earth is distorted from a spherical shape, having an equatorial bulge which gives a component of gravity polewards. This distortion has resulted from the Earth's rotation, and is just sufficient for an equilibrium to be reached so that bodies at rest on the Earth's surface do not experience a net force either polewards or equatorwards. However, a body which is moving to the east at speed u_x is rotating faster than a body at rest, and thus experiences a larger CenF, viz.

$$m\left(\Omega + \frac{u_x}{r}\right)^2 r$$

$$= m\Omega^2 r + \frac{mu_x^2}{r} + m2\Omega u_x .$$

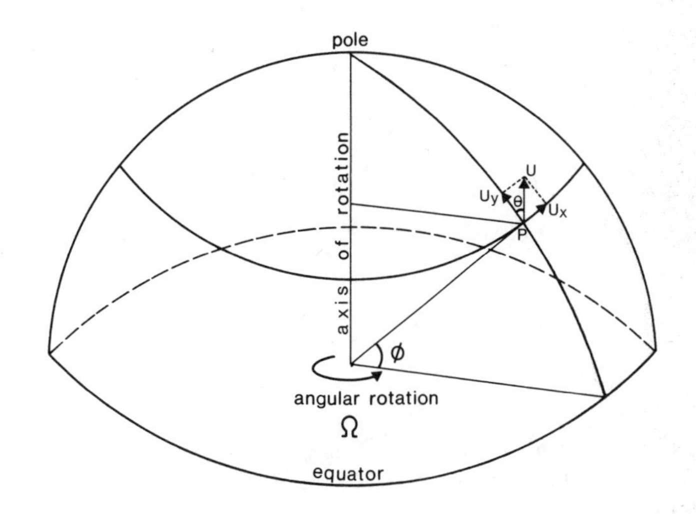

8. 4 Motion on the rotating Earth in the northern hemisphere.

Of the three terms in the expansion on the right-hand side, the first is not due to the movement of the body relative to the Earth, and is balanced by the distortion of the Earth, so it is not a part of the CorF. The second term is due only to the movement of the body and is not dependent on the Earth's rotation, and is thus again not a part of the CorF. In the oceans and atmosphere u_x is small compared with Ωr, its speed due to the Earth's rotation, and thus this second term is very small

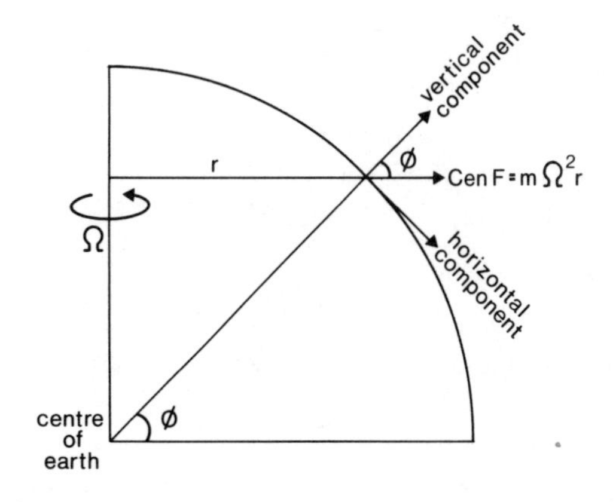

8. 5 The centrifugal force and the Earth's rotation.

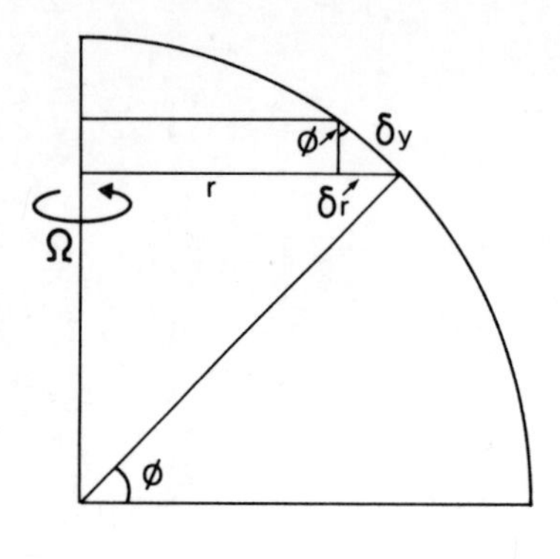

8.6 The change in radius of rotation, δr, as a body moves a distance δy polewards.

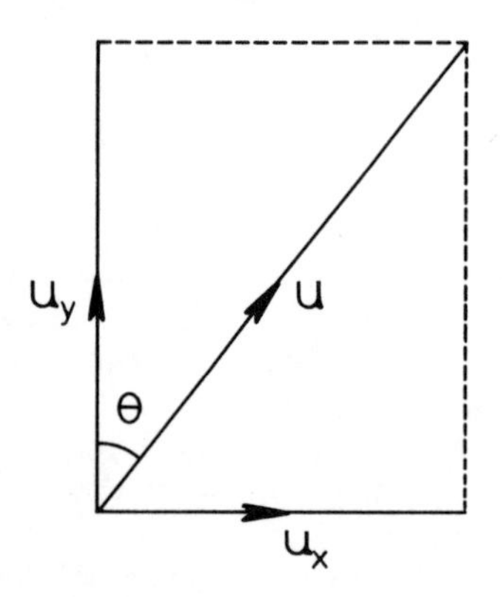

8.7-a The northerly and easterly components of the velocity $\vec{u}$.

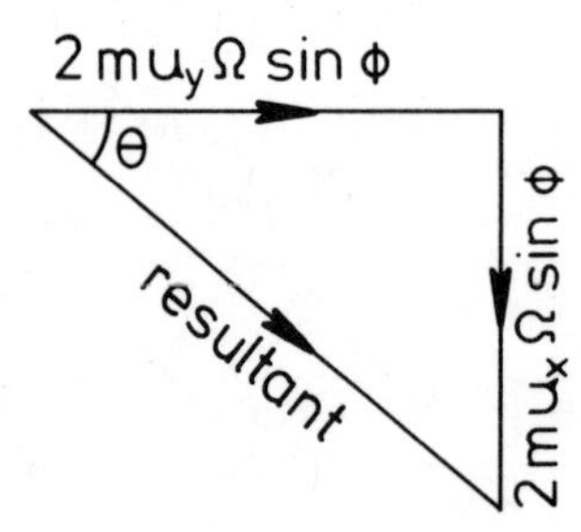

8.7-b The vector addition of the Coriolis Forces resulting from the velocity components shown in Figure 8.7-a.

compared with the third term, $2m\Omega u_x$, which is the CorF. This force is inclined at an angle ϕ from the vertical as in Figure 8.5, and thus its horizontal component is given by $2m\Omega u_x \sin\phi$ and is directed equatorwards. If we had considered u_x to be towards the west, the CenF on the body would have been reduced by its movement relative to the Earth, and the horizontal component of the CorF would thus have been directed in the opposite direction, i.e. polewards. This would be true in the southern hemisphere also, but whereas in the northern hemisphere it means that the CorF is directed 90° to the right of u_x, in the southern hemisphere it is directed 90° to the left of u_x.

The northerly component, u_y: as the body moves horizontally and polewards its radius of rotation decreases, and thus if its velocity remains constant its angular momentum will decrease. The rate of change of angular momentum in this case will be

$$\frac{d}{dt}(m\Omega r^2) = m\Omega 2r\frac{dr}{dt} .$$

Now in a time interval δt, the body moves a distance δy, and its radius of rotation is reduced by δr (Fig. 8.6).

$$\text{As } \delta t \rightarrow 0 , \quad \sin\phi \rightarrow \frac{\delta r}{\delta y} ,$$
$$\text{i.e.} \quad \delta r \rightarrow \delta y \sin\phi .$$

$$\text{Therefore} \quad \frac{dr}{dt} = \lim_{\delta t \to 0} \frac{\delta r}{\delta t} = \lim_{\delta t \to 0} \frac{\delta y \sin\phi}{\delta t} = \frac{dy}{dt}\sin\phi = u_y \sin\phi .$$

Hence the rate of change of angular momentum is $m\Omega 2ru_y\sin\phi$. In the absence of any other force, the body's velocity will change in order to conserve its angular momentum. It will tend to move towards the east in order to increase its angular momentum. This appears to an observer on the Earth's surface to result from a couple, *F.r*, being applied to the body with the point of application of *F* at a distance *r* from the Earth's axis and in the same direction as that of the Earth's rotation:

$$F.r = m\Omega 2ru_y\sin\phi \ ,$$

i.e. F (which is the CorF in this case) $= 2m\Omega u_y\sin\phi$ and is directed towards the east. If u_y had been directed equatorwards, F would have been directed westwards in order to oppose the increase in angular momentum. This means that in the northern hemisphere the CorF is directed 90° to the right of u_y, whilst in the southern hemisphere it is directed 90° to the left of u_y.

Total Coriolis Force: the total horizontal component of the CorF acting on the body as a result of its velocity $\vec{u}$ is given by the vector addition of $2m\Omega u_x\sin\phi$ and $2m\Omega u_y\sin\phi$ as shown in Figure 8. 7. It is $2m\Omega u\sin\phi$, and is directed 90° to the right of $\vec{u}$ in the northern hemisphere, and 90° to the left of $\vec{u}$ in the southern hemisphere. This may be abbreviated by saying 90° *cum sole* of $\vec{u}$, noting that in the northern hemisphere the Sun appears to move across the sky to the right while in the southern hemisphere it appears to move across the sky to the left. The term $2\Omega\sin\phi$ is known as the *Coriolis parameter*, and is commonly represented by the symbol f so that the CorF becomes mfu.

8. 8 Force 9 in the North Atlantic, with spindrift and a short, steep sea running.

9.0 The effects of a point protruding into the sea and of shoaling water on a wave train approaching the shore near Beachy Head on the south coast of Britain.

9 Waves and Tides

Progressive waves: general characteristics, deep water waves and long waves □ Wind waves: generation, main features, shallow water effects □ Tsunamis □ Standing waves: general characteristics □ Tide-generating forces □ The equilibrium tide □ Tidal phenomena in the ocean and in marginal seas: resonance, diurnal and semi-diurnal tides, amphidromic points, tidal currents, tide prediction.

SOME OF THE MOST FAMILIAR natural forms of motion are the waves at the interface between the ocean and the atmosphere. Waves of various types can also exist entirely within either the atmosphere or the ocean. Those involving vertical displacements result from relative motion between adjacent layers of fluid of different density or from some external disturbance such as an earthquake, and in almost all cases depend on gravity as a restoring force to produce an oscillation and are thus known as *gravity waves.* (The exception are capillary waves on the water surface which have lengths of the order of 10^{-2} m and for which surface tension effects become important.) Waves are usually progressive, moving past a fixed point with a regular series of crests and troughs, but if such a wave train encounters a barrier it may be reflected so that the incident and reflected waves become superimposed and a standing wave results. Here we shall consider the wind-generated waves at the ocean surface, and also tides and certain other long waves which are found in the ocean. First, however, there are a number of general features of water waves to be discussed.

PROGRESSIVE WAVES

Any regular, oscillating disturbance which spreads through a medium may be called a *wave.* In a *transverse* wave, the particles of the medium move to and fro at right angles to the direction in which the wave is travelling. Ripples on a pond are an example of transverse wave motion; the direction of propagation is horizontal, and a small particle floating on the surface moves up and down but is ultimately left in its original position after the passage of the wave. It is essentially the *disturbance* which moves forwards, not the water (as long as the wave does not break and there is no wind or friction with the ground). In *longitudinal* waves, the particles of the propagating medium oscillate forwards and backwards in the direction of the wave motion, but here too their mean position does not change. Sound is an example of a longitudinal compression wave. The particles of the medium through their oscillations form alternate regions of compression and rarefaction, and it is the location of these regions which advances, not the medium.

The essential parameters of a wave are its wavelength (L), its amplitude (a) or wave height ($H = 2a$), and its period (T), which is the

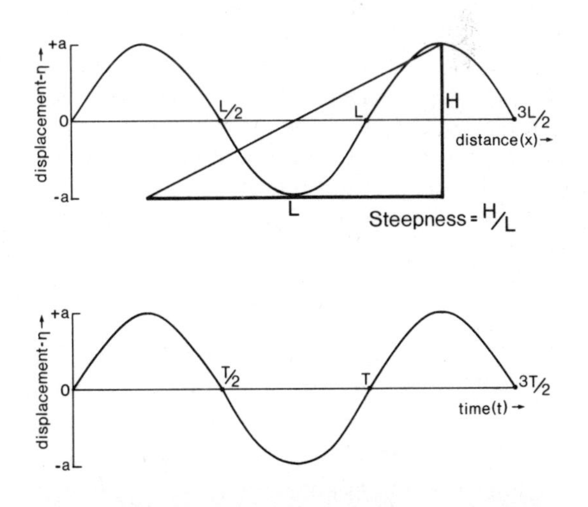

9.1 Displacement plotted against (a) distance, and (b) time, for a sinusoidal wave.

time interval between successive crests (or troughs) passing a fixed point. The number of waves which pass a fixed point in unit time is the wave frequency (f) and is equal to $1/T$. The speed at which the wave is moving (c) is thus given by:

$$c = L/T \;. \qquad (9.\text{I})$$

Wave steepness is defined rather arbitrarily as H/L.

The passage of a wave produces a displacement (η). This may be considered as a function of distance (x) at a fixed time, or as a function of time (t) at a fixed distance. If the wave is assumed to have a sinusoidal profile these are simple harmonic functions, viz.

$$\eta = a \sin \frac{2\pi x}{L} \quad \text{(see Fig. 9. 1-a)}$$

$$\eta = a \sin \frac{2\pi t}{T} \quad \text{(see Fig. 9. 1-b)}$$

or, in general,

$$\eta(x, t) = a \sin\left(\frac{2\pi x}{L} - \frac{2\pi t}{T}\right) . \qquad (9.\text{II})$$

Alternative wave profiles may be assumed, e.g. the trochoid illustrated in Figure 9. 2 gives a better representation of surface water waves which are relatively steep, but the mathematical analysis becomes more difficult in these cases.

It can be shown that, if various simplifying assumptions are made, the speed c of an individual sinusoidal gravity wave of small amplitude on the water surface is related to its wavelength and the depth of water (d) by the expression:

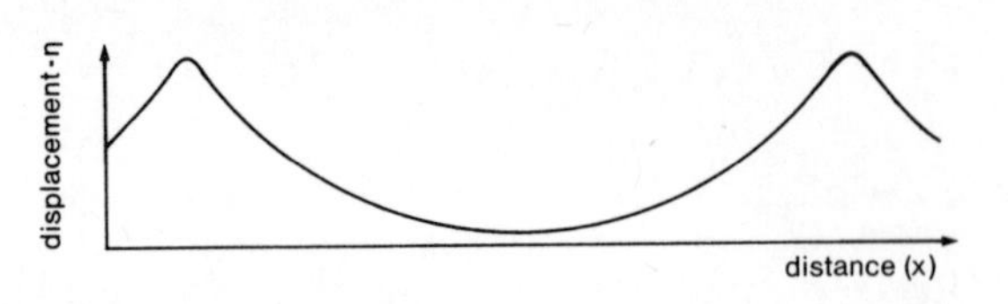

9. 2 A trochoidal wave. A trochoidal curve is the path traced by a point on a circular disc as the disc rolls along a straight line.

$$c = \sqrt{\left(\frac{gL}{2\pi} \times \tanh \frac{2\pi d}{L}\right)} . \qquad (9.\text{III})$$

(Tanh X is a mathematical function known as the hyperbolic tangent. Its important properties here are: (i) when X is small, $\tanh X \simeq X$; (ii) when $X > \pi$, $\tanh X \simeq 1$.) This can be approximated in two particular circumstances as shown in Table 9. 1. In each case the wave speed is dependent on only one variable: the wavelength in the case of a deep water wave, and the water depth in the case of a shallow water wave. For intermediate cases it must be considered to be dependent on both variables.

For a deep water wave we may replace c by L/T, and obtain a relationship between these parameters, viz.

$$L = \frac{g}{2\pi} T^2 . \qquad (9.\text{IV})$$

Thus if T is measured in seconds, $L = 1.56\ T^2$ m and $c = 1.56\ T$ m s^{-1} or $3.03\ T$ knots. (1 knot is 1 nautical mile hr^{-1}, and is widely used in meteorology and oceanography because each minute of latitude is approximately 1 nautical mile, providing a convenient scale on many maps and charts.)

Table 9. 1

	I DEEP WATER, OR SHORT WAVE	II SHALLOW WATER, OR LONG WAVE
Defined by d/L being	$> \frac{1}{2}$	$< \frac{1}{20}$
e.g. (a) If $L = 300$ m, d must be (b) If $d = 4$ km, L must be	$>$ 150 m $<$ 8 km	$<$ 15 m $>$ 80 km
Wave Speed (c) given by	$\sqrt{(gL/2\pi)}$	$\sqrt{(gd)}$

Table 9.2: An abbreviated version of the Beaufort Wind Scale for use at sea

BEAUFORT NO.	NAME	WIND SPEED		STATE OF THE SEA SURFACE	WAVE HEIGHT* (m)
		KNOTS	$m\ s^{-1}$		
0	Calm	<1	0.0–0.2	Sea like a mirror.	0
1	Light air	1–3	0.3–1.5	Ripples with appearance of scales; no foam crests.	0.1–0.2
2	Light breeze	4–6	1.6–3.3	Small wavelets; crests have glassy appearance but do not break.	0.3–0.5
3	Gentle breeze	7–10	3.4–5.4	Large wavelets; crests begin to break; scattered white horses.	0.6–1.0
4	Moderate breeze	11–16	5.5–7.9	Small waves, becoming longer; fairly frequent white horses.	1.5
5	Fresh breeze	17–21	8.0–10.7	Moderate waves taking longer form; many white horses and chance of some spray.	2.0
6	Strong breeze	22–27	10.8–13.8	Large waves forming; white foam crests extensive everywhere and spray probable.	3.5
7	Moderate gale	28–33	13.9–17.1	Sea heaps up and white foam from breaking waves begins to be blown in streaks; spindrift begins to be seen.	5.0
8	Fresh gale	34–40	17.2–20.7	Moderately high waves of greater length; edges of crests break into spindrift; foam is blown in well-marked streaks.	7.5
9	Strong gale	41–47	20.8–24.4	High waves; dense streaks of foam; sea begins to roll; spray may affect visibility.	9.5
10	Whole gale	48–55	24.5–28.4	Very high waves with overhanging crests; sea surface takes on white appearance as foam in great patches is blown in very dense streaks; rolling of sea is heavy and visibility reduced.	12.0
11	Storm	56–64	28.5–32.7	Exceptionally high waves; sea covered with long white patches of foam; small and medium sized ships might be lost to view behind waves for long times; visibility further reduced.	15.0
12	Hurricane	> 64	> 32.7	Air filled with foam and spray; sea completely white with driving spray; visibility greatly reduced.	> 15

*This column gives a rough guide to likely significant wave heights (average of the highest one-third of the waves); for the higher wind speeds it assumes an open ocean situation with the wind velocity having been fairly constant for some hours.

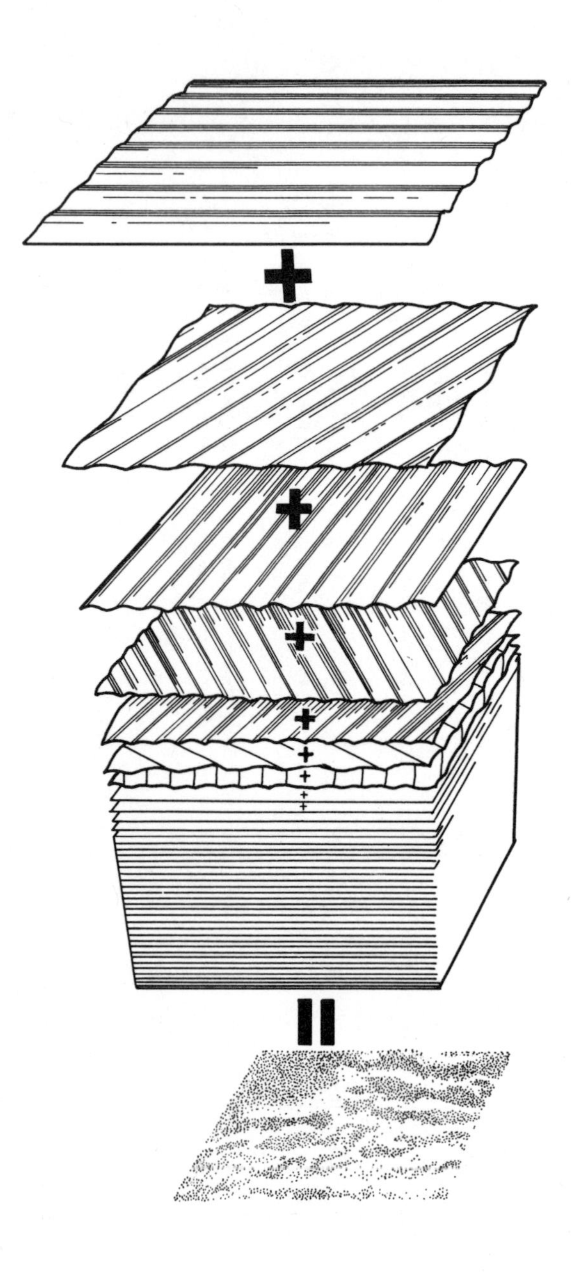

9. 3 The formation of a wave field by the superimposition of a series of simple wave trains moving in different directions. (After W. J. Pierson et al.)

In the areas in which they are generated, wind waves on the ocean surface are almost invariably deep water waves. The actual mechanism by which they are generated has been the subject of much discussion and many theories, and it is still poorly understood. It involves the turbulent fluctuations which are present in the wind, and their interaction with the waves as they move across the water surface, and it transfers considerable quantities of energy from the atmosphere to the ocean. Empirical studies have shown that, if the wind blows at a constant velocity for a sufficient length of time over an adequate length of sea (termed the *fetch*), then a certain wave field, which depends on the wind speed, will be established. This wave field is very irregular, but may be considered to stem from the superimposition of a wide variety of simple wave trains with different periods and lengths (Fig. 9. 3). If the wave field is analysed into its component parts, a wave energy spectrum such as those shown in Figure 9. 4 may be compiled. The energy contained in a wave is related to its height – for a simple harmonic wave it is proportional to the square of the wave height. Thus as the wind speed increases the wave heights in a fully developed sea increase, and it will also be noted from Figure 9. 4 that the period of the waves containing the most energy increases. The relationship between wind speed and the state of the sea surface is used for estimating wind speed at sea by the Beaufort Scale (Table 9. 2). This scale is valid only for waves generated locally, and adequate time and fetch are required for a fully developed sea to become established, e.g. about 10 hr and 100 km for a Beaufort force 5, and 40 hr and 1 000 km for a Beaufort force 8.

Once formed, these waves move away from the area in which they were generated at a speed which is proportional to their period, as shown above. Thus away from a storm area the first waves to arrive will be the longest, and successively shorter waves will follow. In fact the speed at which a *group* of waves advances in deep water is only half the speed of the individual waves, and each individual wave advances through the group but loses height as it approaches the front of the group and then disappears. This effect can be seen by throwing a rock into a large pond and observing the movement of the resultant waves. Observations of the times at which waves of various periods reach a particular location can be used to estimate how far they have travelled since they were generated if their group speed is calculated.

When waves move away from their generating locality their energy becomes spread over a greater area, and is dissipated (particularly in the case of the shortest wavelengths) by viscous effects. The irregular pattern of waves in the generating area thus becomes modified, with wave heights decreasing, short waves disappearing, and smooth long-crested swell remaining. Such swell can travel over very great distances if it remains in deep water – swell arriving on the Alaskan coast has been identified as originating from storms near Antarctica more than 10 000 km away.

PARTICLE MOTION AND SHALLOW WATER EFFECTS

The water particles in deep water waves move in almost circular orbits (Fig. 9. 5-a). The size of the orbits decreases exponentially with increasing depth – at the surface their diameter is given by the wave height, but at a depth of $L/9$ this has been halved, and at a depth of $L/2$ (the minimum water depth for a deep water wave) it is negligible. In fact the circular orbits are not quite closed for waves of finite amplitude because the water moves slightly further forward in the crest than it moves backward in the trough, and this gives a small net displacement of water in the direction of wave travel.

When waves move into water of depth less than $L/2$, the circular orbits of the water particles near the bottom become flattened (Fig. 9. 5-b). The wave speed becomes reduced according to Equation 9. III until, when the depth is $L/20$, it has become equal to $\sqrt{(gd)}$. However, the wave period must remain constant as waves continue to arrive at the same frequency, and thus as the waves slow down their wavelength must also decrease in accord with Equation 9. I. This leads to refraction of the wave pattern, and to the wave crests becoming almost aligned with the submarine contours (Fig. 9. 6).

Wave energy progresses at the group speed of the waves. When they enter shallow water their group speed decreases and so, for the same amount of energy to continue to be transmitted, their height must increase. As their length decreases they must become increasingly steeper, and their particle speeds must increase until the maximum particle speed at the crest exceeds the wave speed. The wave is then unstable, spills over, and breaks.

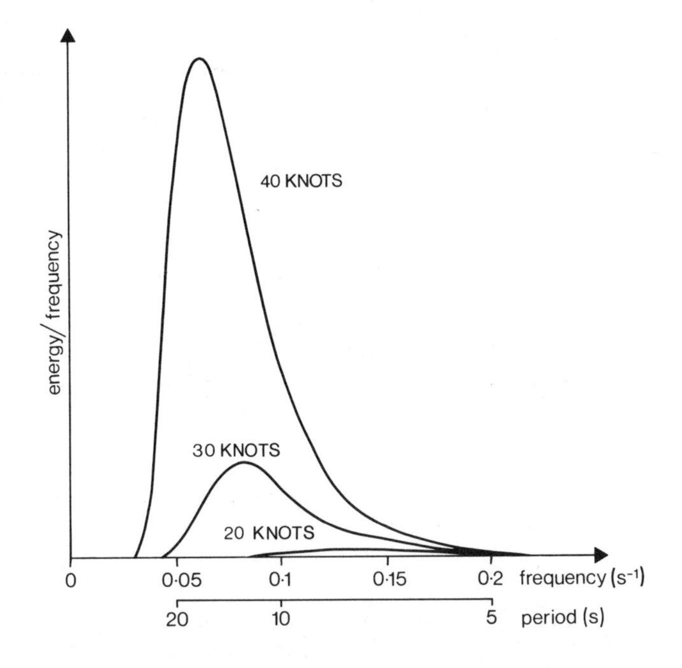

9. 4 Continuous wave spectrum for a fully risen sea at wind speeds of 20, 30 and 40 knots. (After W. J. Pierson et al.)

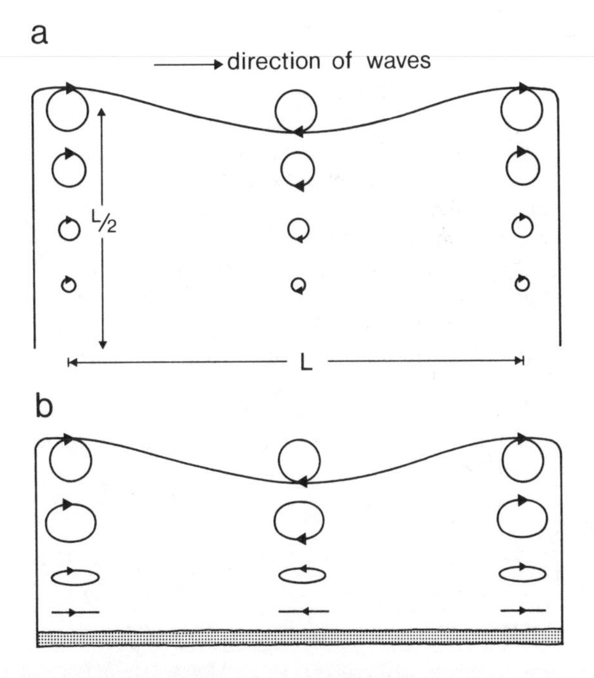

9. 5 Particle motion in (a) a deep-water wave, and (b) a shallow-water wave.

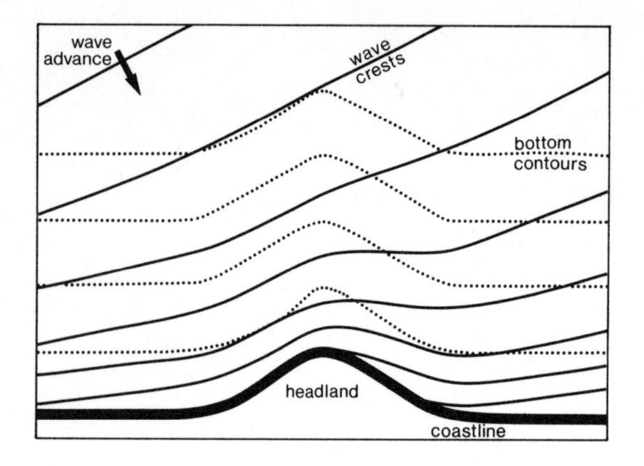

9.6 The refraction of waves advancing obliquely towards a coastline.

It is here, on the coastline, that most of the energy which the waves obtained from the wind is either used in eroding and transporting beach and cliff material, or is dissipated as heat. Refraction generally concentrates wave energy on headlands, and thus wave erosion tends to produce a uniform coastline.

TSUNAMIS

The waves which are generated by disturbances of the ocean floor, as when an earthquake occurs, have been called *seismic* sea waves and (erroneously) tidal waves, but here the Japanese name *tsunami* is used. These waves radiate outwards from a seismic disturbance and have very long periods and wavelengths – typically 15 to 20 minutes and 200 km. Hence they behave as long waves, and their speed, given by $\sqrt{(gd)}$, is some 200 m s^{-1} or 400 knots for open ocean depths.

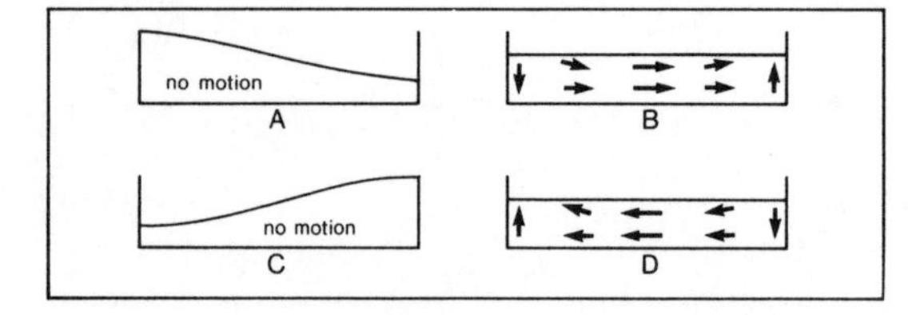

9.7 Standing wave in a tank.

In the open ocean their wave heights are usually less than 1 m and so they pass ships completely unnoticed. When they approach shallow water, however, their speed is reduced and their height increases dramatically, so that at a coastline they can form breakers of enormous magnitude which rush ashore and flood areas up to 10 m or more above normal sea level.

The nature of tsunamis varies considerably, presumably with the seismic events which generate them. Some comprise a single crest, but in other cases a broad trough precedes the main crest, or there may be a series of crests. They are essentially features of the Pacific Ocean, where seismic activity is associated with the relative movements of the plates shown in Figure 1.2, and a tsunami warning system is in operation in the Pacific area. The pattern of destruction which occurs at a particular stretch of coast depends on the nature of the tsunami, and on the sea-bed contours in the vicinity of the location, which determine the way in which it is refracted. Some tsunamis have caused considerable loss of life, whilst many cause virtually no damage at all, and the major problem for the warning system lies in distinguishing between these.

STANDING WAVES

Standing waves, known as *seiches*, can occur in closed basins such as lakes, and in bays which are open to the sea at one end. Those in closed basins can readily be modelled in a laboratory by setting the water in a tank into oscillation (Fig. 9.7). At either end of the tank (the antinodes) the water level is alternately high and low, whilst in the middle of the tank (the node) the water level remains constant but horizontal motions are greatest. In this case the length of the tank or basin, l, corresponds to half the length of the wave (L). If d/l is less than 1/10, as will be the case in most natural situations, this can be considered as a shallow water wave, and so

$$\frac{2l}{T} = \sqrt{(gd)}\,,$$

i.e.

$$T = \frac{2l}{\sqrt{(gd)}}\,. \qquad (9.\text{V})$$

Thus the period of the seiche in the basin is determined by the length of the basin and the depth of water in it, and constitutes its *resonant* period. In

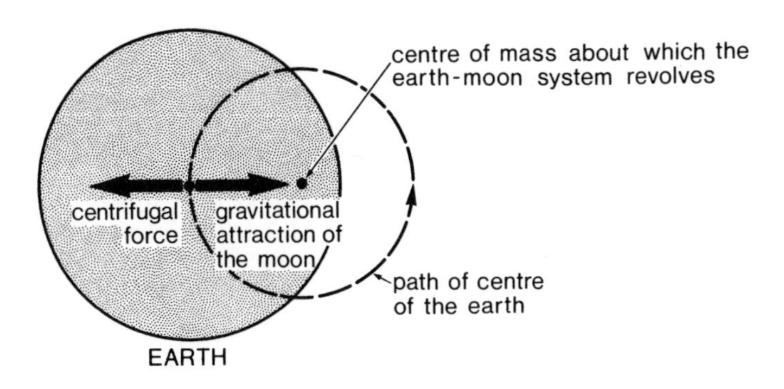

9. 8 The balance of the gravitational attraction and centrifugal force for the Earth-Moon system.

a bay which is open at one end, a seiche can occur with a node at the mouth and an antinode at the head. The length of the bay (l) then corresponds to only $L/4$, and its resonant period is given by:

$$T = \frac{4l}{\sqrt{(gd)}} . \qquad (9.\ \mathrm{VI})$$

In both cases it is also possible for seiches to be established with more than one node and with correspondingly shorter periods.

As an example, consider a harbour, open to the sea, which is 3 km long and has a mean depth of 40 m. Its resonant period, according to Equation 9. VI, will be 10 min. If the water in the harbour is disturbed, perhaps by a storm or by a wave motion in the sea outside with a period of this order, a harbour seiche can develop and a phenomenon known as *surging* occurs which brings difficulties to moored vessels.

THE EQUILIBRIUM TIDE

Tides result from the differential gravitational attractions of the Moon and Sun over the surface of the Earth. For the Earth as a whole the gravitational attraction of the Moon just balances the centrifugal force which results from the rotation of the Earth about the centre of mass of the Earth-Moon system (Fig. 9. 8). This centrifugal force is the same for a unit mass at any point on the Earth's surface, but the gravitational attraction varies according to the distance of the unit mass from the Moon. There is thus a net force directed towards the Moon at the point on the Earth's surface closest to the Moon, and a net force directed away from the Moon at the point on the Earth's surface furthest from the Moon. At other positions on the Earth's surface there are smaller net forces directed at some angle to the vertical (Fig. 9. 9). If the ocean surface were to reach equilibrium with these forces it would have to become perpendicular everywhere to the resultant of this force and the Earth's gravitational attraction. Such an *equilibrium tide* would be essentially an ellipsoid, with its bulges where the gravitational attraction of the Moon was greatest and least.

Such a situation is never achieved because of the rotation of the Earth. To retain the shape of the equilibrium tide the oceanic bulges would have to travel round the Earth at such a speed that they remained in the same position relative to the Moon. This means that they would have to travel right round the Earth in 24 hr 50 min, the period of the Earth's rotation about its axis *with respect to the Moon*. This is greater than 24 hr because the Moon is rotating about the Earth-Moon centre of mass, in the same direction as the Earth rotates, with a period of 27.3 days. Even if

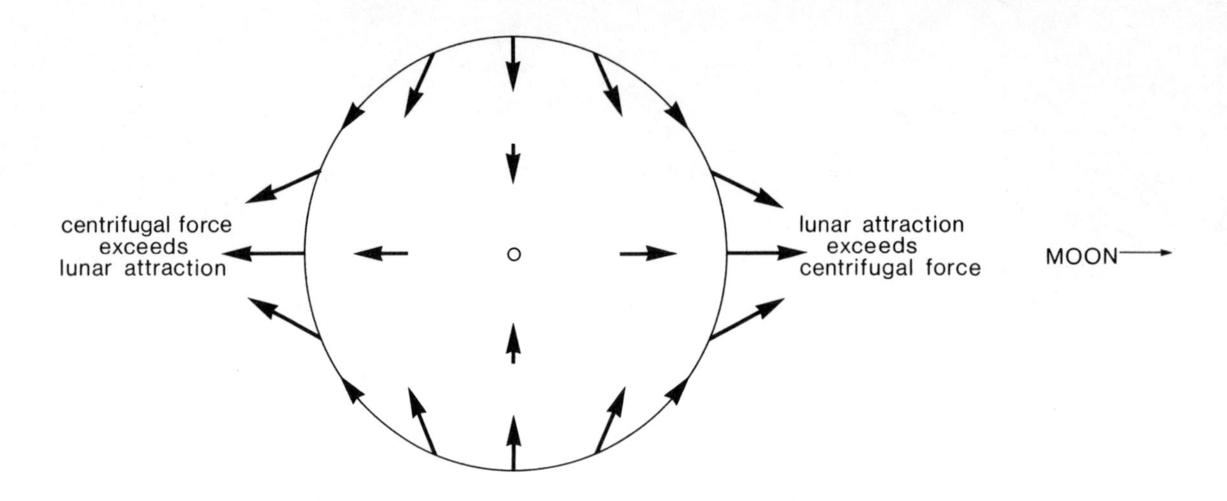

9. 9 The pattern of the main tide-producing forces over the Earth's surface.

the Earth were entirely covered by water this would be possible only if the depth of water permitted the bulges, travelling as long waves at a speed given by $\sqrt{(gd)}$, to keep up with the Moon. This would require a depth of about 20 km near the equator.

The equilibrium tide concept does, however, enable us to understand the periodicity of the tides. Essentially we have two wave crests in every 24 hr 50 min, and thus the basic lunar tidal period is the semi-diurnal one of 12 hr 25 min.

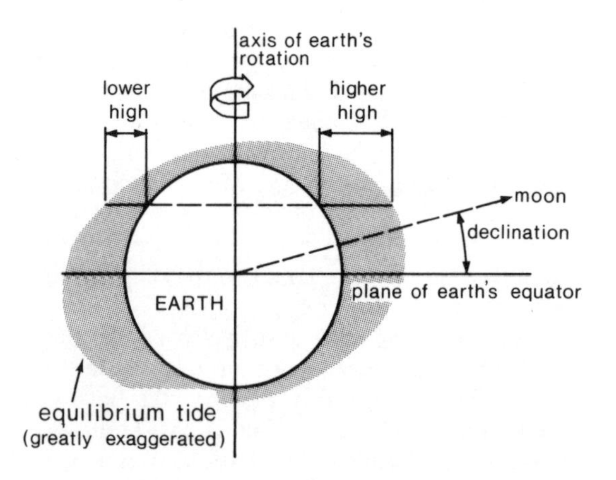

9. 10 The effect of the Moon's declination on the equilibrium tide. (After J. Williams et al.)

These two crests are not however of equal magnitude unless the Moon is directly above the equator. The declination of the Moon (Fig. 9. 10) varies during a month and may be up to $28\frac{1}{2}°$. This produces a diurnal inequality in the tide, and we may consider the actual tide as comprising simple harmonic constituents with periods of 12 hr 25 min and 24 hr 50 min which have been superimposed (Fig. 9. 11).

The Sun also exerts tide-producing forces at the Earth's surface. As the magnitude of these forces is directly proportional to the mass of the attracting body, but inversely to the cube of its distance away, the solar equilibrium tide is only 46% as great as that of the Moon. It provides tidal constituents with periods of 12 hr and 24 hr. When these are superimposed on the lunar constituents, we find that we get maximum tides known as *spring tides* every 14 days at new and full Moon when the Sun and Moon act together, and minimum or *neap tides* between, when they are acting against one another (Fig. 9. 12). Again the declination of the Sun changes, and other regular variations in the relative positions of the Sun, Earth and Moon, including their distances apart, produce very many harmonic constituents in the tide.

OBSERVED TIDES

Observations of water level around the shores of the oceans reveal variations which display the semi-diurnal and diurnal periods and the spring-

9.11 A tidal curve with diurnal inequality (A), and its diurnal (B) and semi-diurnal (C) constituents.

neap cycle which the equilibrium tide would lead us to expect. They are, however, rarely in phase with the equilibrium tide, and their amplitudes vary considerably from one place to another. At Lowestoft, on the east coast of England, high water occurs some $9\frac{1}{2}$ hr after and 3 hr before the transit of the Moon, and the mean tidal range is 1.5 m, whereas at Harwich, only 70 km away, high water occurs about 2 hr later than at Lowestoft, and the mean tidal range is 3.0 m.

The effect of the tide-producing forces is to set up waves in oceans, seas and bays which have periods that coincide with those of the major tidal constituents. In most places the semi-diurnal period is the dominant one, but in some oceanic areas the diurnal period becomes dominant. It is, in fact, the horizontal components of the tide-generating forces, known as the *tractive forces*, which are essentially responsible for producing these waves. In some cases they behave as progressive waves, travelling across ocean basins at speeds determined by the depths of the basin. If the resonant period of the basin is close to that of a tidal constituent, a seiche with that period is also established. When the progressive waves encounter coastlines, and in particular when they enter marginal seas and bays, they are subject to reflection, and if there is resonance then tidal oscillations of considerable amplitude are experienced.

The most notable example of such resonance is provided by the Bay of Fundy in eastern Canada. This bay is about 270 km long and has an average depth of about 70 m, and if we substitute these values in Equation 9. VI we find its resonant period to be about 12 hr. It is not surprising, therefore, that very large semi-diurnal tidal oscillations are experienced at its head, the range exceeding 15 m on spring tides.

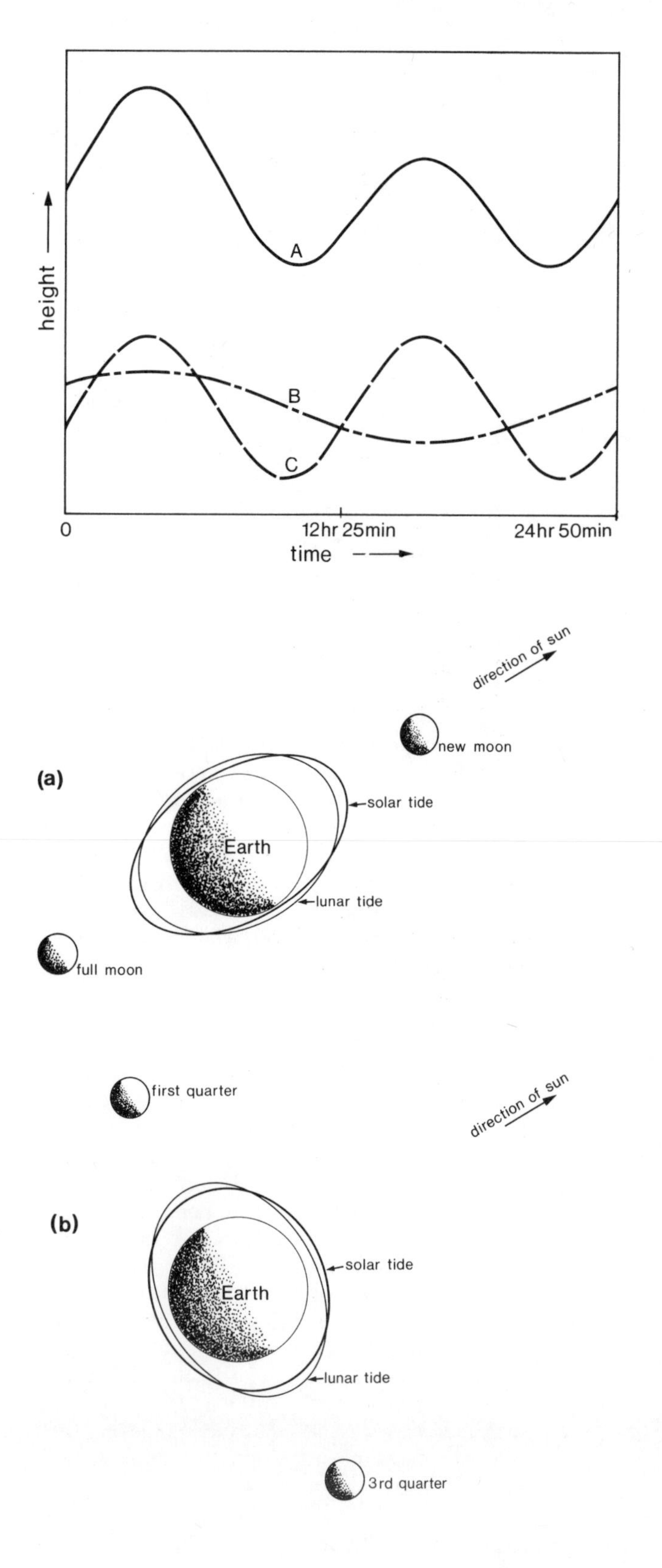

9.12 Diagrammatic representation of the solar tide superimposed on the lunar tide to give (a) spring tides, and (b) neap tides.

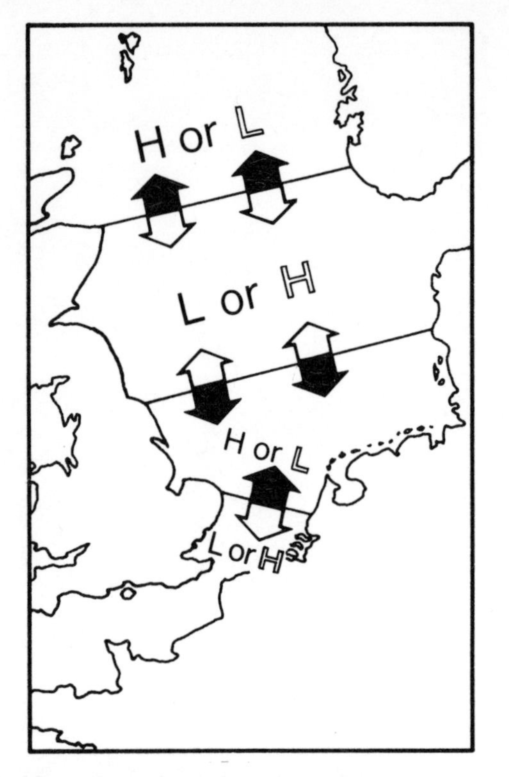

9. 13 A hypothetical multi-nodal seiche in the North Sea. The open arrows show the currents which would flow across the nodal lines during one half of the seiche, and the open letters indicate whether Low or High water would follow these at each of the antinodes. The solid arrows indicate the currents during the other half of the seiche, and the solid letters the levels which would follow these at the antinodes.

The North Sea has a resonant period of the order of 40 hr – the actual value depends on the axis which is used to determine its length and depth. Progressive tidal waves enter it from the adjacent Atlantic Ocean to the north, and also through the restricted Straits of Dover. A multi-nodal seiche with three nodal lines (Fig. 9. 13) would have a period close to the semi-diurnal tidal period, and this is essentially the tidal system which develops in the North Sea, but it is modified by the CorF. As the water flows across the nodal lines it is deflected *cum sole*. Hence, instead of simply oscillating to and fro, the high water circulates round a point on the nodal line known as an *amphidromic point*. The tidal range increases with distance outwards from the amphidromic point as shown in Figure 9. 14.

As in other waves, there are horizontal movements of the water particles associated with the vertical oscillations. These *tidal currents* have the same periodicities as the vertical tidal oscillations, and (as in the case of other long waves) their velocities increase when the wave enters shallower water. On the continental shelves tidal currents typically reach speeds of 0.5 to 1.0 m s^{-1} (1 to 2 knots). In restricted channels they flow in one direction for one half of the tidal cycle and then reverse and flow in the opposite direction for the other half of the cycle. If the tidal wave is a standing wave the current reversals will occur at low and high water, but if it is a progressive wave they will occur at mid-tide. In wide bays and the open sea the CorF causes the tidal current to change direction constantly, and the path which the water particles describe in a horizontal plane is typically an elliptical one.

Tides are essentially oceanic phenomena. Atmospheric tides do exist, and lead to regular variations of atmospheric pressure at the Earth's surface, but the ranges of those with periods related to the lunar cycle are extremely small (about 0.2 mb or less), and are concealed by much larger variations from other causes so that they can only be detected when long records are subjected to sophisticated methods of analysis. A regular oscillation of pressure does, however, occur with the solar semi-diurnal period of 12 hr. This oscillation is readily observed in low latitudes where its range is greatest (2–4 mb) and where fewer other variations obscure it. Pressure maxima in this oscillation occur at about 1000 and 2200 hr local time. The extent to which this phenomenon results from tidal forces or from solar heating has not yet been satisfactorily resolved.

9. 14 (Opposite) The amphidromic tidal system of the North Sea. The co-tidal lines (continuous) show the time of high water in 'lunar hours' (approximately 1 hr 2 min) after the Moon's meridian passage at Greenwich, and the co-range lines (dotted) show the average tidal range.

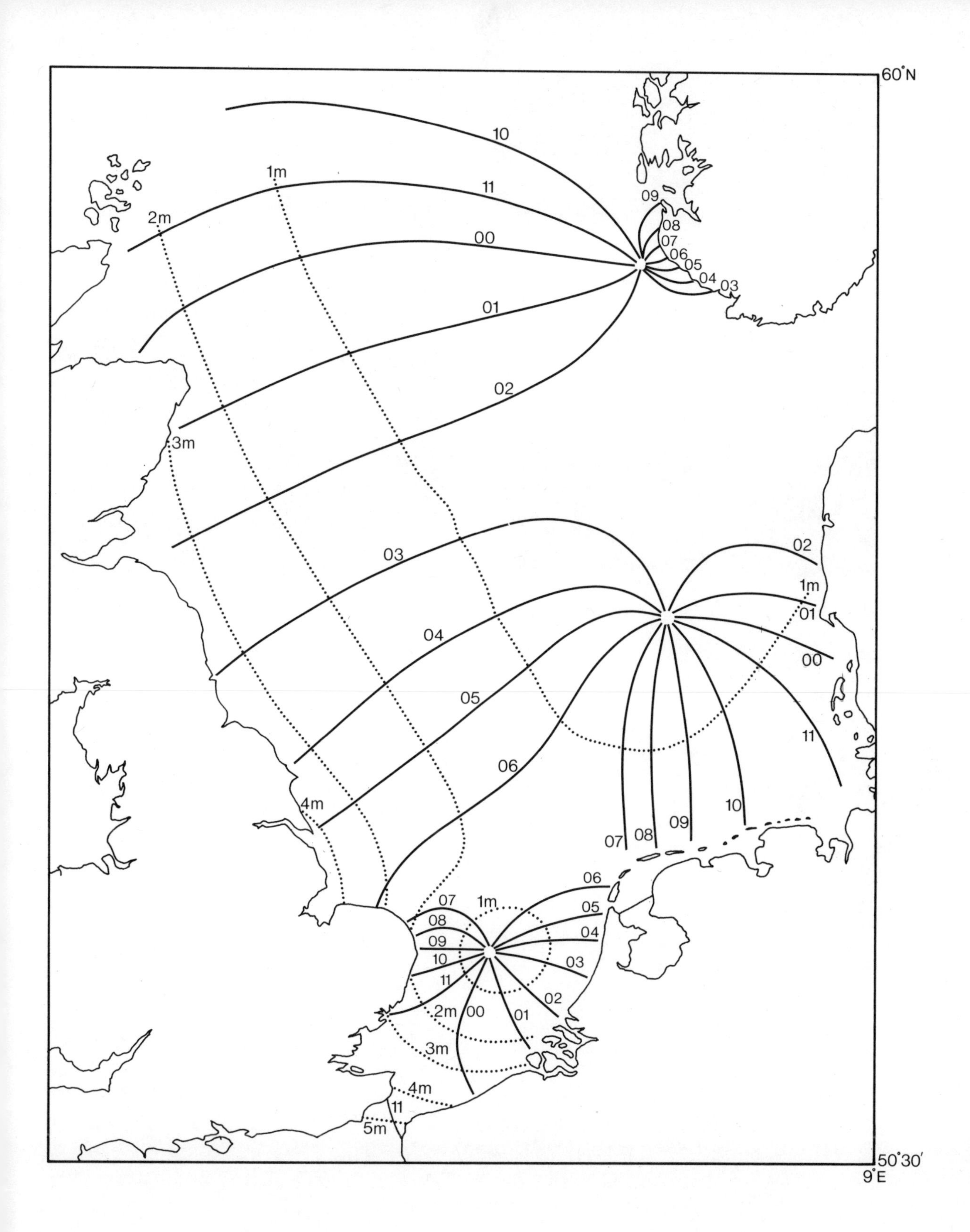
60°N
10
1m
11
2m
09
08
00
07
06
05
04
03
01
02
3m
02
03
1m
01
04
00
05
11
06
4m
10
07
08
09
06
07
1m
05
08
04
09
10
03
11
02
2m
00
01
3m
4m
11
5m
50°30′
9°E

10.0–a A water spout in the Atlantic, photographed from M.V. British Consul at a distance of ca. 6–9 km. Note the relative calm of the sea in the foreground.

10.0–b Water spout at San Feliu de Guixols, Spain, in the afternoon of September 2, 1965.

10 Pressure Gradients and Associated Winds

Land and sea breezes □ Horizontal pressure gradients and associated force □ Anabatic and katabatic winds □ Geostrophic, surface, gradient and cyclostrophic winds.

THE FORCES WHICH ARE MOST frequently responsible for the initiation and maintenance of larger-scale motions in the atmosphere and ocean are the horizontal pressure gradients which result from the uneven distribution of mass within the fluids. The pressure may vary along a horizontal surface due to variations in the height or the density of the fluid above it. A horizontal surface is simply defined as being perpendicular to the direction of apparent gravity as indicated by a plumb-line responding to the resultant of gravitational attraction and the CenF from the Earth's rotation. A surface which is everywhere at the same pressure is known as an *isobaric* surface, and any line on this surface is an *isobar*. In this chapter we first consider winds which are initiated by pressure gradient forces due to the uneven heating of the Earth's surface, and then go on to see how these are modified on a larger scale by secondary forces.

LAND AND SEA BREEZES

Consider two columns of air *A* and *B* of equal cross-sectional area and initially having the same density and pressure at every level. Column *A* is located over the land, and column *B* over the adjacent sea. During the daytime the temperature of the land surface will rise higher than that of the sea surface (as explained in Chapter 5), and this will lead to the air in column *A* being warmed more than that in column *B* and thus expanding. Some of the expansion will be lateral, so that the total mass of air remaining in column *A* will be reduced and the pressure at its base fall below that at the base of *B*; but some of the expansion will be upwards, so that *above* a certain level (perhaps 1 km) the mass of air in column *A* will *exceed* that in column *B* and the horizontal pressure gradient will be from *A* to *B* (Fig. 10.1-a). A horizontal pressure gradient established at either

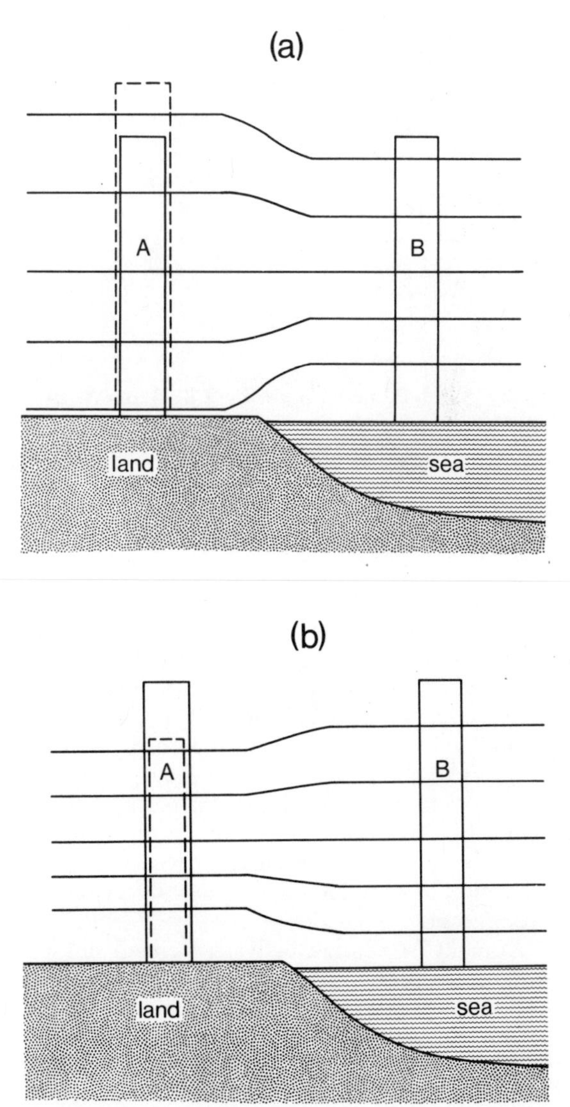

10.1 The relative expansion and contraction of columns of air over land and sea, and the associated isobaric patterns (a) during the day, and (b) at night.

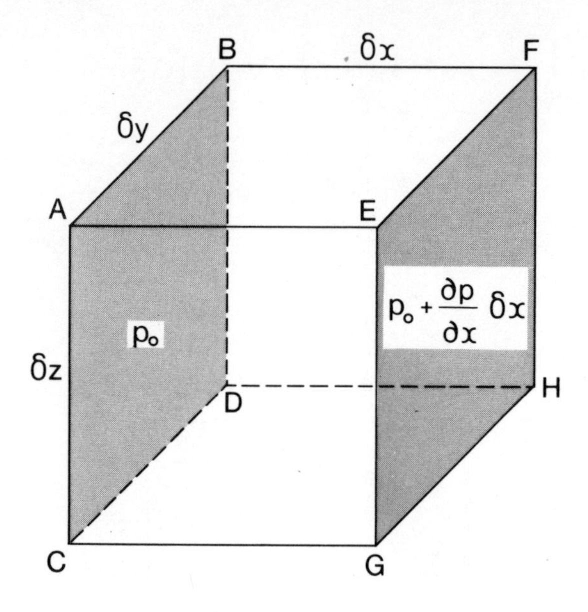

10. 2 A cuboid acted upon by a horizontal pressure gradient $\partial p/\partial x$.

level will be sufficient to initiate a circulation cell with a *sea breeze* blowing near the surface (i.e. a breeze blowing *from* the sea towards the land).

Such sea breezes are most likely to be established when the wind is otherwise light and the sky is cloudless so that the maximum contrast between sea and land develops. They increase in strength during the day to reach perhaps 10 m s^{-1} by late afternoon, and also extend over a greater area as the day proceeds, penetrating perhaps 50 km inland by evening. They bring cool and moist air to the land, where instability often leads to the formation of cumulus cloud in the advancing sea breeze front.

At night the contrast between the land and sea surface temperatures is reversed, and thus the horizontal pressure gradients and the circulation cell are also reversed (Fig. 10. 1-b). The surface wind is now a *land breeze*, which is usually very much lighter than the daytime sea breeze – perhaps 2–3 m s^{-1}. Again it is likely to be established during clear nights with little wind from other sources, the conditions which favour the formation of radiation fog (Chap. 3), and land breezes may thus carry radiation fog some distance out over the sea.

When land and sea breezes start to blow, secondary forces will begin to act on the air. In low latitudes the CorF, being a function of latitude, is unlikely to play any significant rôle in motions on the time and distance scales of land and sea breezes, and an approximate balance is established between the pressure gradient force and friction. In higher latitudes, however, the CorF may become important, and will act to turn the wind less directly across the coastline.

HORIZONTAL PRESSURE GRADIENT FORCE

To determine the magnitude of the pressure gradient force, consider a small cuboid of fluid with dimensions δx, δy, δz (Fig. 10. 2). Let the pressure at the face *ABCD* be p_o, and the pressure gradient *along* the x-axis (horizontal) be $\partial p/\partial x$. The force which the cuboid experiences as a result of the pressure on one of its faces is given by the product of the pressure and the surface area of that face. The resultant force on the cuboid due to the difference in the pressures at the faces *ABCD* and *EFGH* will thus be

$$p_o\delta y\,\delta z \;-\; \left(p_o + \frac{\partial p}{\partial x}\delta x\right)\delta y\,\delta z$$
$$= -\frac{\partial p}{\partial x}\delta x\,\delta y\,\delta z\ .$$

$\delta x\delta y\delta z$ is the volume of the cuboid. If its density is ρ and its mass m, its volume will be m/ρ and thus the pressure gradient force will be given by:

$$\mathrm{PGF} = -\frac{m}{\rho}\frac{\partial p}{\partial x}\ . \qquad (10.\,\mathrm{I})$$

For example, a cuboid of air near sea level with sides 1 m and density 1.2 kg m^{-3} will have a mass of 1.2 kg. If there is a horizontal pressure gradient of 1 mb in 100 km, the resultant force on the cuboid will be $\frac{1}{100}\times 10^{-3}$ mb m^2 = 10^{-3} N .

KATABATIC AND ANABATIC WINDS

Contrasts in the heating and cooling of air, similar to those which cause land and sea breezes, are also encountered in mountainous areas. During the night the highest land will cool down most, causing the air above it to contract, and a wind system similar to that of land breezes is established. In this case, however, the slope of

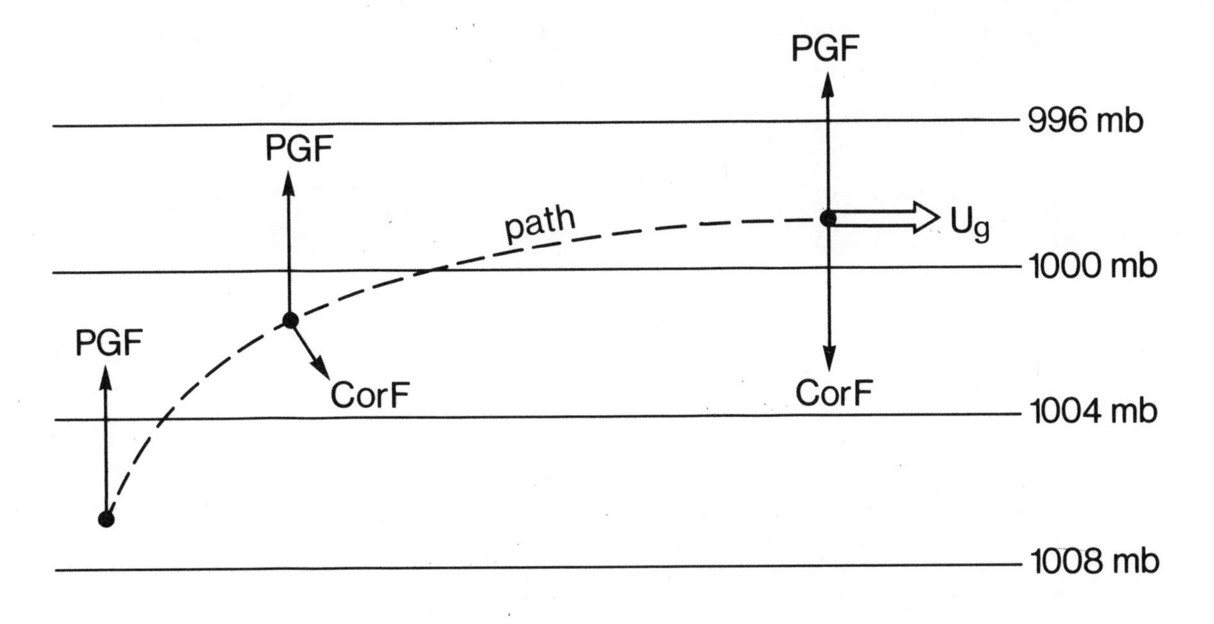

10. 3 The establishment of the geostrophic wind (northern hemisphere).

the ground will lead to a down-hill gravitational flow of the colder, denser air making the wind much stronger than a land breeze. These *katabatic winds* can be particularly violent when they originate over a snow- or ice-covered surface and are confined by narrow glacial valleys and fjords, as around Greenland. Katabatic winds, or winds with a katabatic component, are frequently experienced in certain areas and are given local names. In the Adriatic Sea a katabatic wind from between north and east which usually occurs in winter is known as the *Bora*. The *Mistral*, which blows down the Rhone Valley and over the Gulf of Lions, is associated with a horizontal pressure gradient from north-west to south-east, but it is strengthened considerably by the katabatic flow of air down the mountains. Where such winds reach the sea, being cold and dry they increase the density of water through cooling and evaporation, and thus promote convective mixing of the water in addition to mechanically-induced turbulence.

During sunny days in summer the higher parts of mountains, and particularly the slopes facing the Sun, can become appreciably warmer than adjacent valleys. This produces a circulation cell similar to that of a sea breeze, with weak up-slope *anabatic winds*, and with convection clouds above the mountain summits and ridges.

GEOSTROPHIC WINDS

Once a wind is initiated by a horizontal pressure gradient, secondary forces will begin to act. When the wind is well-established, and providing it is some way from the equator, the CorF is usually the most important of these. Consider a 'parcel' of air moving from an area of high pressure to one of low pressure in the northern hemisphere as a result of the pressure gradient, with the isobars running in straight lines, and in the absence of friction (Fig. 10. 3). The CorF will act so as to turn it to the right, and the resultant of the pressure gradient force (PGF) and the CorF will cause the parcel to move faster. As it speeds up, the CorF, being proportional to u (the speed of the parcel), will increase and turn it increasingly to the right. At the point where it is moving perpendicular to the PGF, the CorF and the PGF act in opposite directions, and the resultant force will depend on which of them is stronger. If it is the PGF the wind will experience an acceleration to the left, its speed will increase, CorF will increase and the parcel swing back to the right. If it is the CorF which is stronger it will be deflected further to the right, its speed will decrease, CorF will decrease and it will swing back to the left. Equilibrium may eventually be achieved if the PGF remains constant with the wind blowing perpen-

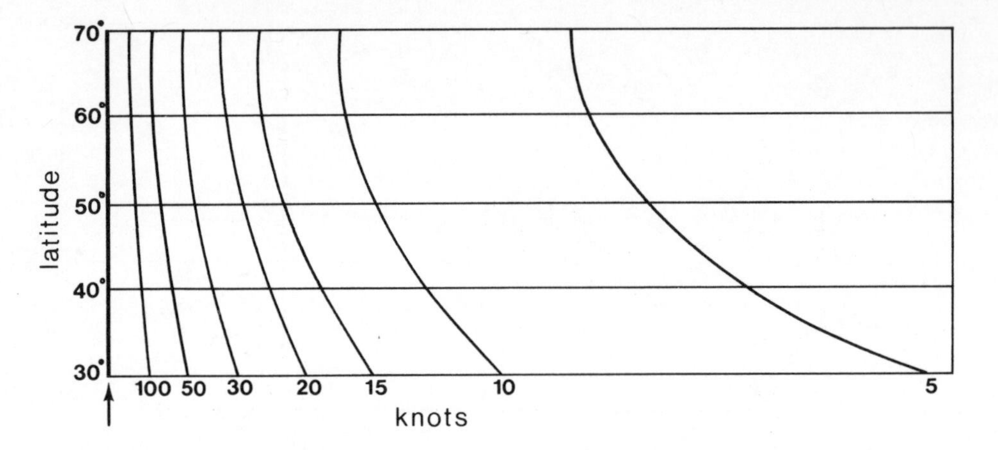

10. 4 A geostrophic wind scale for use with a map on a conical orthomorphic projection with standard parallels 30° and 60°, where the scale is 1:15 million. This scale gives wind speed in knots when used for the spacing of isobars at 4 mb intervals, and air at 1 000 mb pressure and temperature 15°C.

dicular to the PGF, the PGF and the CorF being exactly equal and opposite so that the parcel of air experiences no acceleration. This is known as a *geostrophic* wind. It blows parallel to the isobars, with the area of high pressure on its right in the northern hemisphere. In the southern hemisphere the high pressure area is on its left. This is in agreement with Buys-Ballot's Law which was formulated in the 19th century from observations of the wind and which states: *In north latitudes, face the wind and the barometer will be lowest to your right. In south latitudes, face the wind and the barometer will be lowest to your left.*

The speed of the geostrophic wind, U_g, can be determined by equating the magnitudes of the PGF and the CorF, viz.

$$\frac{m}{\rho}\frac{\partial p}{\partial x} = 2mU_g\Omega\sin\phi \, ,$$

i.e.

$$U_g = \frac{1}{2\Omega\sin\phi\rho}\frac{\partial p}{\partial x} \, . \qquad (10.\ \mathrm{II})$$

E.g. with a pressure gradient of 1 mb in 100 km in latitude $\phi = 30°$, taking Ω as 2π rad in 23 hr 56 min (the *sidereal* day) $= 7.29\times10^{-5}$ s^{-1}, and ρ as 1.2 kg m^{-3},

$$U_g = \frac{10^{-3}\ \mathrm{kg\ s^{-2}\ m^{-2}}}{2\times7.29\times10^{-5}\times0.5\times1.2\ \mathrm{kg\ s^{-1}\ m^{-3}}}$$

$$\simeq 11.4\ \mathrm{m\ s^{-1}} \, .$$

(The sidereal day is the time required for the Earth to rotate through 360° relative to distant stars. Due to the Earth's movement in its orbit round the Sun, the civil day is approximately 4 minutes longer.)

On synoptic weather maps showing pressure distribution by means of isobars, a geostrophic wind scale is provided for determining U_g from the spacing of the isobars. A standard spacing of 4 mb is usual. As the relationship between U_g and the horizontal pressure gradient varies with latitude, the scale must be graduated for specific latitudes, and allowances can also be made for variations in air density when necessary. Figure 10. 4 gives an example of such a scale.

In deriving the geostrophic wind, we made two important assumptions: that the isobars ran in straight lines, and that friction was absent. Where these assumptions are not valid, or close to the equator where the CorF becomes zero, the actual wind differs from the geostrophic wind. We shall consider below examples of such situations.

SURFACE WIND

At the Earth's surface friction is important, slowing the wind down. This retardation induces a vertical shear in the wind in the lowest 500 m or so of the atmosphere, and the consequent friction makes the wind deviate significantly from the geostrophic wind up to about this level. Close to the Earth's surface, ignoring the drag of the overlying air, there are three forces which must balance for a steady state to be achieved, as shown in Figure 10. 5. The magnitude and direction of the PGF are fixed, but the magnitudes and directions of the CorF and friction vary according to the wind speed and direction. To obtain a balance the wind direction must be inclined at some angle α to the isobars, and the magnitude of the CorF must be less than that of the PGF. The speed of the surface wind is thus less than that of the geostrophic wind.

The amount by which it is reduced and the magnitude of α depend upon the nature of the surface over which the wind blows, as this, together with the wind speed, determines the magnitude of the frictional force. The standard height at which surface wind measurements are made is 10 m, and at this height over the sea the wind speed is about two-thirds U_g and α is usually between 10° and 20°. Over rough ground the wind speed at 10 m may be no more than one-third U_g, and α may be 40° or more. Such allowances can be made to estimate the surface wind when the geostrophic wind has been determined from an isobaric chart.

As one ascends in the atmosphere the frictional force is caused by the differences between the drags of the overlying air and the underlying air, and is not strictly in the opposite direction to that of the wind velocity. However, similar considerations to the above apply, and as the frictional force becomes smaller with increasing height, the wind speed and direction gradually become closer to those of the geostrophic wind.

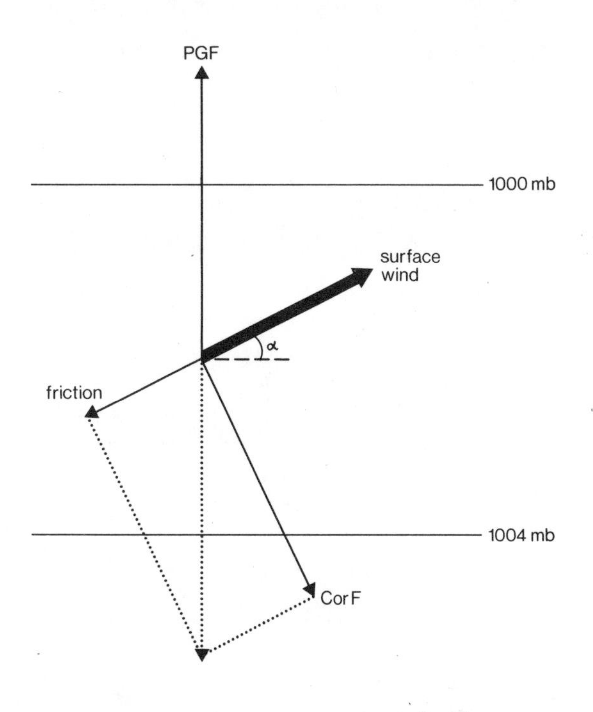

10. 5 Balance of forces in a surface wind (northern hemisphere).

GRADIENT WIND

When the isobars are curved, the geostrophic wind would follow a curved path on the Earth's surface and thus be subject to a centrifugal force (CenF) in addition to the other forces considered above. The two alternative situations which can arise with curved isobars are the *cyclonic* situation in which pressure decreases towards the centre of curvature (Fig. 10. 6-a) and the *anticyclonic* one in which pressure increases towards the centre of curvature (Fig. 10. 6-b). In each case, if the wind follows the isobars the CenF will act outwards from the centre of curvature. Thus in the cyclonic situation, where the wind turns anticlockwise in the northern hemisphere and clockwise in the southern hemisphere, the CenF will act in the same direction as the CorF, and so a smaller CorF (and hence a smaller wind speed) is required to obtain a balance with the PGF. This wind is known as the *gradient* wind, U_{gr}, and its magnitude can be obtained from the equation:

$$\frac{m}{\rho}\frac{\partial p}{\partial x} = 2\,mU_{gr}\Omega\sin\phi + \frac{m(U_{gr})^2}{r},$$

$$\text{i.e.}\quad U_{gr} = U_g - \frac{(U_{gr})^2}{2\Omega r\sin\phi}. \qquad (10.\text{ III. a})$$

In each hemisphere, the wind direction in an anticyclonic situation is opposite to that in a cyclonic situation, the CenF is opposed to the CorF and thus the CorF must be increased by a higher wind speed than in the geostrophic case. The gradient wind speed for an anticyclonic situation is thus obtained from:

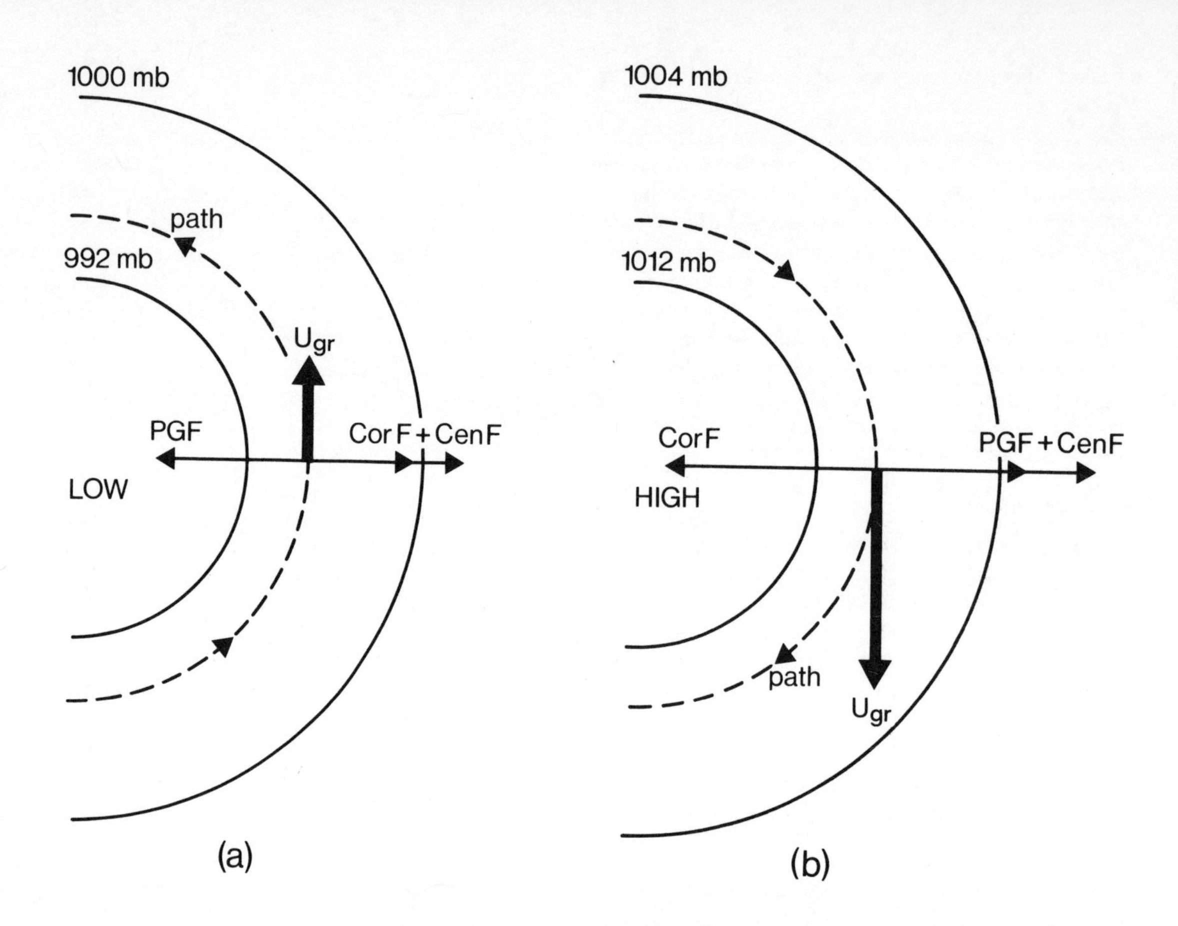

10. 6 Balance of forces and establishment of the gradient wind (a) for a cyclone, and (b) for an anticyclone

$$\frac{m}{\rho}\frac{\partial p}{\partial x} + \frac{m(U_{gr})^2}{r} = 2mU_{gr}\Omega\sin\phi ,$$

i.e.
$$U_{gr} = U_g + \frac{(U_{gr})^2}{2\Omega r\sin\phi} . \quad (10. \text{ III. b})$$

If, in the example used above to calculate the geostrophic wind speed in latitude 30°N with $\partial p/\partial x = \frac{1}{100}$ mb km^{-1}, the isobars had had a cyclonic curvature with radius 1 000 km, the gradient wind speed would have been 10.0 m s^{-1}, but had they had anticyclonic curvature of the same radius it would have been 14.2 m s^{-1}. There is, however, an upper limit to the gradient wind speed in the anticyclonic situation above which no balance can be achieved between the PGF, the CorF and the CenF because the CenF increases as the square of the wind speed. This upper limit is given by $\Omega r\sin\phi$, and if the wind speed exceeds this there will be a net force outwards from the centre of curvature, and the wind will blow across the isobars, reducing the magnitude of the PGF.

The gradient wind thus blows in the same direction as the geostrophic wind, but at a different speed. Near the ground, however, the gradient wind will be affected by friction in just the same way as the geostrophic wind, its speed being reduced and its direction becoming inclined across the isobars.

CYCLOSTROPHIC WIND

In a vortex of small radius with a large PGF, the CorF becomes small in comparison with the other forces which are involved, and in the limiting case the PGF and the CenF balance one another. We may then write

$$\frac{m}{\rho}\frac{\partial p}{\partial x} = \frac{m(U_{cyc})^2}{r},$$

i.e. the *cyclostrophic wind speed* is

$$U_{cyc} = \sqrt{\left(\frac{r}{\rho} \cdot \frac{\partial p}{\partial x}\right)} . \qquad (10.\text{IV})$$

This gives a close approximation to the wind speed which is encountered in a tornado or in the corresponding but less violent feature which originates over the ocean, a water spout. These storms have diameters of the order of 100–200 m, and the air at the surface spirals inwards across isobars due to friction, and then ascends at the low pressure centre carrying up dust and spray. They are associated with extreme instability, and are characterised by funnel-shaped clouds projecting down from a cumulonimbus cloud to the surface. Usually they are very short-lived, lasting perhaps 15 minutes, but their cyclostrophic winds may reach 100 m s^{-1} or more and these, together with the extremely low pressure at their centre (perhaps 25 mb below that in the surrounding air), can cause considerable damage.

In deriving the cyclostrophic wind speed we have assumed that the CorF is not significant, in which case the wind can rotate in either direction about the centre. But as a rule the CorF has sufficient influence, at least during the early stages in the formation of a tornado, to ensure that the wind circulation is cyclonic, so that the CorF is directed outwards.

11.0 A T.S.D. gauge being hoisted back on board a research ship. Such a probe, when lowered into the ocean, provides a continuous record of Temperature and Salinity against Depth. It is usually sent down together with a water-sampling bottle equipped with precision reversing thermometers for calibration purposes.

More sophisticated versions of the device can also monitor such additional parameters as dissolved oxygen concentration, pH, turbidity, velocity of sound and concentrations of specific ions.

11 Geostrophic Currents and Thermal Winds

Geostrophic currents and their relationship to the slope of the sea surface and to density distribution □ Thermal winds.

HORIZONTAL PRESSURE GRADIENTS also exist within the oceans, and associated with them are geostrophic currents. Typically these gradients are one or two orders of magnitude greater than those found in the atmosphere, but as the density of water is almost a thousand times greater than that of air at sea level, the speeds of geostrophic currents are usually between one and two orders of magnitude less than those of geostrophic winds. At the base of the atmosphere a horizontal level can be fairly easily established, and pressure measurements, made with mercury or aneroid barometers, can be corrected to this level given a knowledge of air temperature. The correction amounts to about 1 mb for every 10 m, and an accuracy of pressure at the selected level of about ± 0.1 mb can usually be achieved. In the ocean, however, it is less easy to fix a horizontal datum level precisely, particularly when there are waves on the sea surface and the sea surface itself may be sloping, and an error of 0.1 m in depth relative to a horizontal surface in the ocean will cause pressure to differ by about 10 mb. It is thus impossible to use direct measurements of horizontal pressure gradients to determine geostrophic currents, and instead they must be deduced from the factors which are responsible for the gradient.

There are three principal causes of horizontal pressure gradients in the oceans: a horizontal pressure gradient in the overlying atmosphere; a slope of the sea surface relative to a horizontal surface; and horizontal variations in water density in the ocean. The first of these, as noted above, usually accounts for something between 1% and 10% of the actual horizontal pressure gradient, and thus in many determinations of geostrophic currents it is ignored.

GRADIENT EQUATION

If atmospheric pressure is constant along the sea surface which slopes by a small angle θ, and if we assume that a balance is achieved between the force which results from this slope and the horizontal component of the CorF resulting from the movement of the water, so that a steady state situation exists, we find (Fig. 11. 1) that, writing u for the speed of the geostrophic current, f for the Coriolis parameter ($2\Omega\sin\phi$) and g for the acceleration due to gravity,

$$mg\tan\theta = mfu .$$

i.e.

$$u = \frac{g\tan\theta}{f} , \qquad (11.\ \mathrm{I.\ a})$$

and, as with the geostrophic wind, this current is directed 90° *cum sole* of the downslope direction.

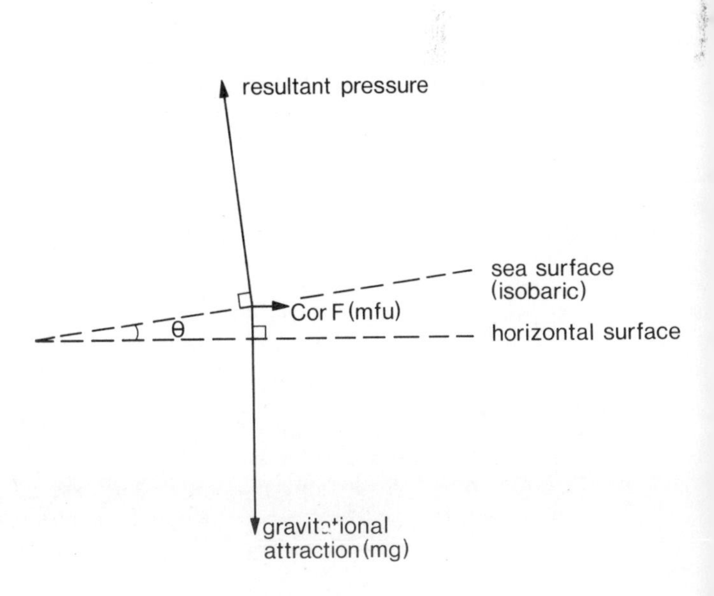

11. 1 The forces acting on water which is moving as a result of a sloping sea surface. The necessary condition for these forces to balance is that mfu/mg = tanθ .

Alternatively we may write

$$\tan\theta = \frac{fu}{g}, \qquad (11.\text{I. b})$$

which is known as the *gradient equation*. Here θ is the angle between an isobaric and a horizontal surface, and if no horizontal variations in water density exist, the isobaric surfaces are inclined to the horizontal by this angle at all depths. Such conditions are called *barotropic* conditions and are characterised by *isopycnic* surfaces (i.e. surfaces of constant density) following isobaric surfaces, and the geostrophic current remaining constant with depth.

Such conditions may be considered to exist within the mixed waters of the Straits of Dover. The mean geostrophic current above the level of frictional influence from the sea-bed in 1957 and 1958 was found to be approximately 0.2 m s^{-1} towards the east. Using Equation 11. I. b, and taking g as 9.81 m s^{-2} and the latitude ϕ as 51°,

$$f = 2\times 7.29\times 10^{-5}\times 0.78 \text{ s}^{-1},$$

$$\text{i.e.} \quad \tan\theta = \frac{2\times 7.29\times 10^{-5}\times 0.78\times 0.2 \text{ m s}^{-2}}{9.81 \text{ m s}^{-2}}$$

$$= 2.3\times 10^{-6}.$$

Taking the breadth of the Straits of Dover as 35 km, this slope means that the mean sea level was some 0.08 m higher on the French side than on the English side in association with the geostrophic current, a fact of importance to engineers for any Channel Tunnel project. The above example shows that it is not always appropriate to think of a horizontal pressure gradient as the cause of a geostrophic current. The flow through the Straits of Dover is rather the result of conditions in the adjacent parts of the North Atlantic, the English Channel and the North Sea, and may be considered to be the cause of the sea surface sloping downwards from France to England in order that a geostrophic balance may be established.

HELLAND-HANSEN'S EQUATION

Where horizontal temperature and salinity variations exist in the ocean they lead to baroclinic conditions in which isopycnic and isobaric surfaces intersect. In such cases geostrophic currents vary with depth, and these variations can be calculated from a knowledge of the density distribution. But to determine the actual current at any one depth requires that the slope of the isobaric surface or the value of the current at some level, a reference level, be known. This is usually a fairly deep level, but currents at surface or near surface level may be used to deduce currents at greater depths. In the simplest case the isobaric surface is horizontal at the reference level so that the current is zero there, and that is the case which will be considered here.

In the situation illustrated in Figure 11. 2 the isobaric surface is known to be horizontal at the reference level, depth z_o. Observations of temperature and salinity have been made at positions A and B, and these are to be used to calculate U_1, the component of the geostrophic current perpendicular to the line AB, at depth z_1. The mean density ρ_A of the water in column A between depths z_1 and z_c, is greater than ρ_B, in the same depth range in column B. The hydrostatic equation assumes that a fluid is at rest, but in the ocean, where water movements are slow and almost horizontal, it gives a sufficiently close approximation to the relationship between pressure and mass. With this assumption made, the isobaric surface at depth z_1 must slope down from B to A, and we will denote this slope by θ_1. Let the height of the isobaric surface p_1 above the reference level be h_A in column A and h_B in column B, and let us take AB as the distance apart of columns A and B.

$$\text{Then} \quad \tan\theta_1 = \frac{h_B - h_A}{AB},$$

and from Equation 11. I. a

$$U_1 = \frac{g(h_B - h_A)}{f \,.\, AB}.$$

From the hydrostatic equation (Equ. 2.I)

$$h_A \rho_A g = h_B \rho_B g,$$

$$\text{i.e.} \quad h_A = h_B \frac{\rho_B}{\rho_A}.$$

$$\text{Thus} \quad U_1 = \frac{g h_B\left(1 - \frac{\rho_B}{\rho_A}\right)}{f \,.\, AB}. \qquad (11.\text{II})$$

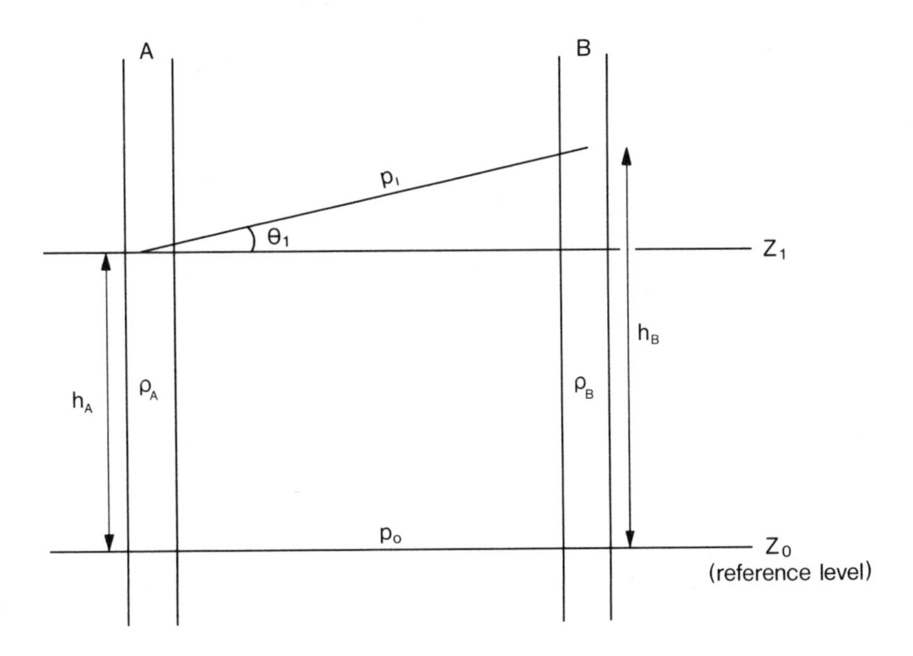

11.2 Horizontal (z) and isobaric (p) surfaces in relation to two vertical columns of water, A and B, with densities ρ_A and ρ_B. ($\rho_A > \rho_B$)

This is a form of Helland-Hansen's Equation which has been used since the beginning of this century for the calculation of geostrophic currents. (For the derivation of Helland-Hansen's Equation in the more general case, and details of the way it is usually applied in oceanography, see Proudman: *Dynamical Oceanography*, Methuen 1953.)

In order to apply the equation in this form an approximate value must be used for h_B, but given that $(z_o - z_1)$ is usually of the order of 10^2 m and that h_B differs from this by <1 m, it is adequate to use the value of $(z_o - z_1)$. For example, if $z_o = 2\ 000$ m, $z_1 = 1\ 000$ m, $\rho_A = 1.0348 \times 10^3$ kg m^{-3}, $\rho_B = 1.0346 \times 10^3$ kg m^{-3}, $\phi = 30°$, and $AB = 50$ km, the component of the geostrophic current at 1 000 m which is perpendicular to AB,

$$U_1 = \frac{9.8 \text{ m s}^{-2} \times 1\ 000 \text{ m} \left(1 - \dfrac{1.0346}{1.0348}\right)}{2 \times 7.29 \times 10^{-5} \times 0.5 \text{ s}^{-1} \times 50 \times 10^3 \text{ m}}$$

$$\simeq 0.39 \text{ m s}^{-1},$$

and the difference $h_B - h_A$ is approximately 0.19 m.

The geostrophic current is directed 90° *cum sole* from the direction in which the isobaric surface slopes downwards. Thus if a deep reference level is being used and the geostrophic currents are being calculated from data obtained above this, they will be directed so that in the northern hemisphere the denser water is on the left looking in the direction of flow, and in the southern hemisphere the denser water is on the right. If, however, a surface or near-surface reference level is used, and the geostrophic currents are calculated downwards from this, the situation is reversed.

For this method of calculating geostrophic currents to be valid, the assumptions for geostrophic flow must hold, i.e. the only forces acting on the water must be the PGF and the CorF, and these must be in balance so that the water experiences no acceleration. In practice the water in the ocean experiences other forces from friction and from wind stress at the surface, and both accelerations and vertical motion disturb the equilibrium. Further, it is only the component of the geostrophic current perpendicular to the line joining the positions at which density has been determined that are obtained, and to establish the complete geostrophic current field the density must be determined at a grid of stations rather than just along a line. The greatest problem, however, in using this method in oceanography is in establishing a reference level. Much effort has gone into determining reference levels of no motion from water mass analysis, variation of dissolved oxygen concentration with depth,

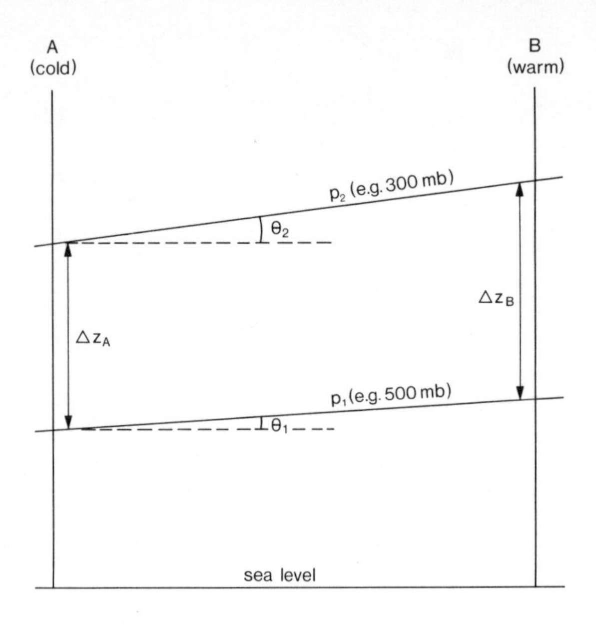

11. 3 Schematic relationship between isobaric layer thickness (Δz) and temperature in the atmosphere.

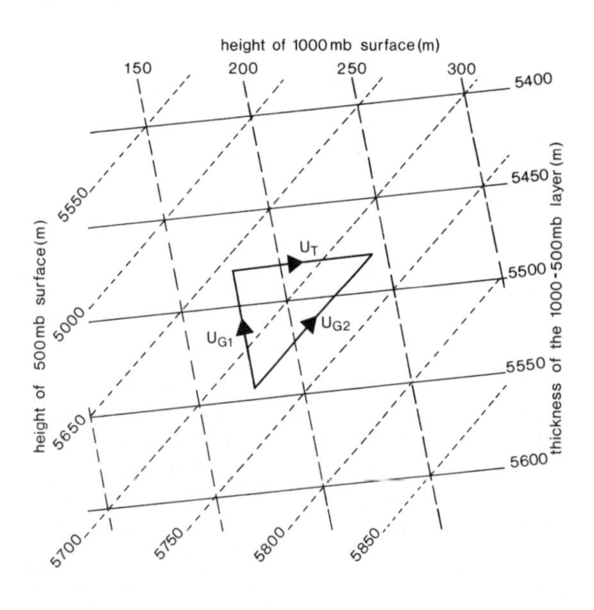

11. 4 The vector addition of the geostrophic wind at 1 000 mb (U_{G1}) and the thermal wind related to the thickness of the 1 000 mb – 500 mb layer (U_T), to obtain the geostrophic wind at 500 mb (U_{G2}).

salinity distribution, and by several other means, but no generally satisfactory method has been found. The increasing availability of direct current measurements probably holds out the best prospects, but these data are showing considerable variation and bring the existence of geostrophic equilibrium in the oceans into question. The other possible source of major error is in the determination of the density field – the calculated currents are proportional to the ratio of density values which typically differ by 0.1 kg m^{-3} or less, and thus random errors of 1 in 10^5 in density determinations will seriously affect the accuracy of the deduced values for the geostrophic current. Moreover, the density values should be obtained simultaneously, and where the density field is being disturbed by any periodic motion such as the tide, mean values of density should be obtained over the period of this motion at each location. Such procedures are rarely possible in practice.

THERMAL WINDS

A method similar to that of Helland-Hansen's Equation can be used to calculate the wind shear in the atmosphere, again assuming that there are no frictional forces. It enables one to determine the winds in the upper atmosphere without making direct pressure measurements there. As in the ocean, the change in geostrophic flow with height depends on the change in slope of the isobaric surfaces, and thus on the horizontal change in thickness of the layer between two isobaric surfaces (Fig. 11. 3).

The components of the geostrophic winds at pressures p_1 and p_2 perpendicular to AB are given by

$$U_{G1} = (g \tan\theta_1)/f \quad \text{and}$$

$$U_{G2} = (g \tan\theta_2)/f ,$$

and so

$$U_{G2} - U_{G1} = g(\tan\theta_2 - \tan\theta_1)/f = \frac{g(\Delta z_A - \Delta z_B)}{f . AB},$$

where Δz is the thickness of the layer between pressures p_2 and p_1.

The thickness of an isobaric layer in the atmosphere may be determined from the finite difference form of Equation 2. V:

$$\Delta z = \frac{R'\overline{T}}{g} \cdot \frac{\Delta p}{p}$$

or $\Delta z = K\overline{T}$, where $\overline{T}$ is the mean temperature of the layer and K is a constant for any two particular isobaric surfaces.

Thus $U_{G2} - U_{G1}$

$$= \frac{g}{f} \cdot \frac{K(\overline{T}_A - \overline{T}_B)}{AB} \qquad (11.\text{III})$$

and is directly proportional to the gradient of layer mean temperature. It is therefore known as the *thermal wind*, and its direction is parallel to the mean isotherms (or thickness contours), with the cold air on its left side in the northern hemisphere and on its right side in the southern hemisphere.

In practice U_{G1} is taken as the geostrophic wind at 1 000 mb computed from direct observations of atmospheric pressure. The thermal wind, which is deduced from charts showing layer thickness (e.g. between 1 000 and 500 mb) compiled from upper-air temperature data, is then used to calculate the geostrophic wind at the top of the layer, U_{G2}. The thermal wind usually blows in a direction other than that of the geostrophic wind at 1 000 mb, and thus must be added vectorially to the latter, as shown in the example in Figure 11. 4.

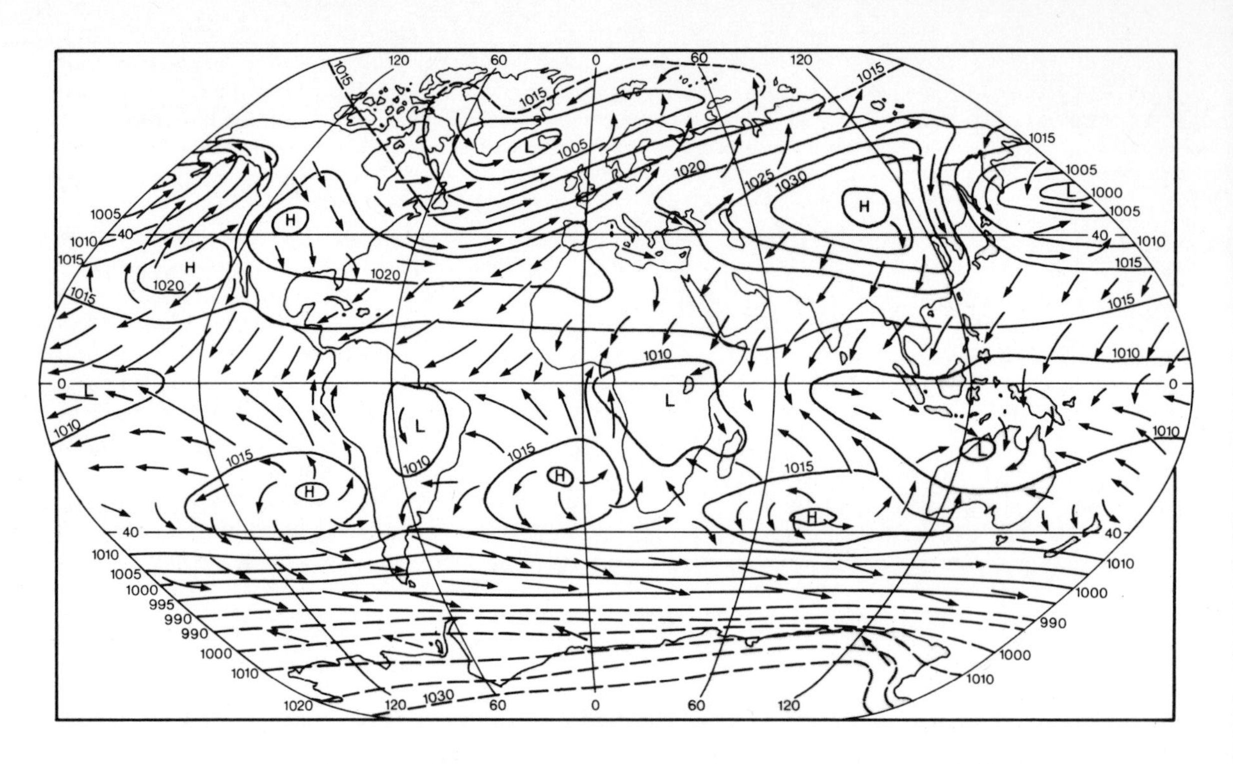

12. 2-a The average distribution of pressure reduced to sea level, and associated wind field in January.

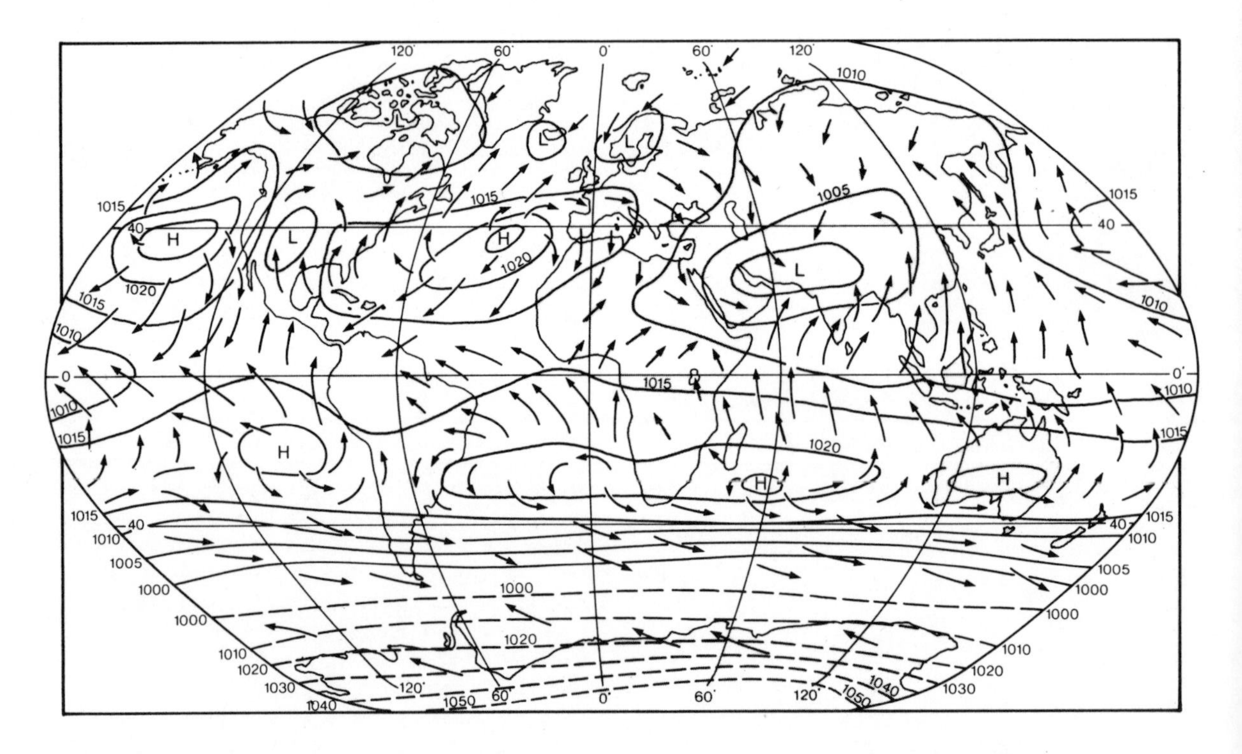

12. 2-b The average distribution of pressure reduced to sea level, and associated wind field in July.

12 Atmospheric Circulation

Pattern on a uniform, non-rotating globe □ Actual surface pressure and wind patterns □ Effect of Earth's rotation □ Dishpan experiments □ The Hadley cell □ The upper westerlies, jet streams and Rossby waves □ The mid-latitude westerlies □ Influence of continents □ Monsoons.

JUST AS LAND AND SEA BREEZES result from horizontal pressure gradients caused by uneven temperature distribution over land and sea, the major planetary wind belts result from the uneven temperature distribution between low and high latitudes. If the Earth did not rotate, and if its surface were entirely uniform as regards albedo, transparency to solar radiation, heat capacity and thermal conductivity, then we might expect a simple convection cell circulation to exist in the troposphere in each hemisphere (Fig. 12. 1). Each cell would, however, have a horizontal dimension of the order of 10^4 km compared with a vertical dimension of only some 10 km. There could then be interference between the upper poleward flow and the lower equatorward flow, leading perhaps to the break-up of the convection cell into a number of smaller cells. When the rotation of the Earth and the non-uniformity of its surface are introduced, the situation becomes considerably more complex.

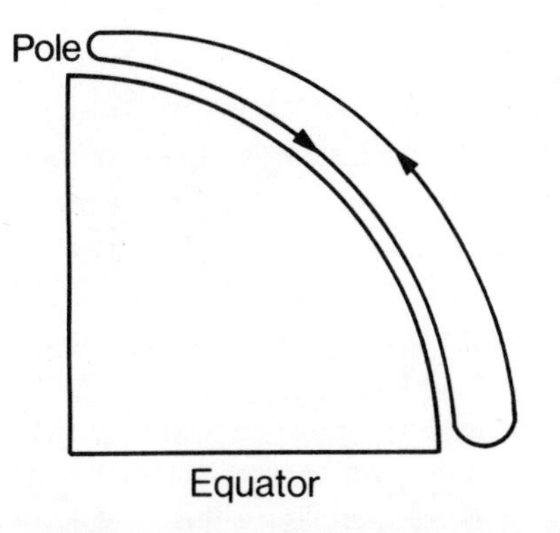

12. 1 Convection cell circulation on a non-rotating uniform Earth.

SURFACE PRESSURE AND WIND PATTERNS

Figure 12. 2 shows the average distribution of atmospheric pressure reduced to sea level for the months of January and July, and the associated surface winds. If mean values are taken for each latitude to remove the zonal differences caused by the distribution of continents and oceans, a series of zonal pressure belts and wind systems is revealed (Fig. 12. 3) which might be taken to represent the mean pattern which would exist if the Earth's surface were uniform. The mean wind fields over the Atlantic and Pacific oceans bear quite a close resemblance to this idealised pattern. Whereas the equatorial low pressure belt and the high pressure at the poles are expected from the simple convection cell illustrated in Figure 12. 1, the sub-tropical high pressure belt and the low pressure belt at about 60° latitude are unexpected. They must, therefore, be dynamic rather than thermal in origin, and the mid-latitude winds associated with the pressure gradient between them are thermally indirect.

It may be noted here that if there were a pressure gradient extending all the way from the poles to the equator, the surface wind on the rotating Earth would have an easterly component in all latitudes, and would therefore be opposed to the Earth's rotation over the whole globe. The atmosphere and the Earth are frictionally coupled to one another, and if the rate of rotation of the Earth is not to be changed by the atmospheric circulation there must be a balance between winds with easterly and westerly components at the Earth's surface so that no net frictional torque is imparted. But, for the reasons stated in Chapter 5, heat must be transported polewards in all latitudes. Two mechanisms may be proposed to achieve this in the thermally indirect wind system

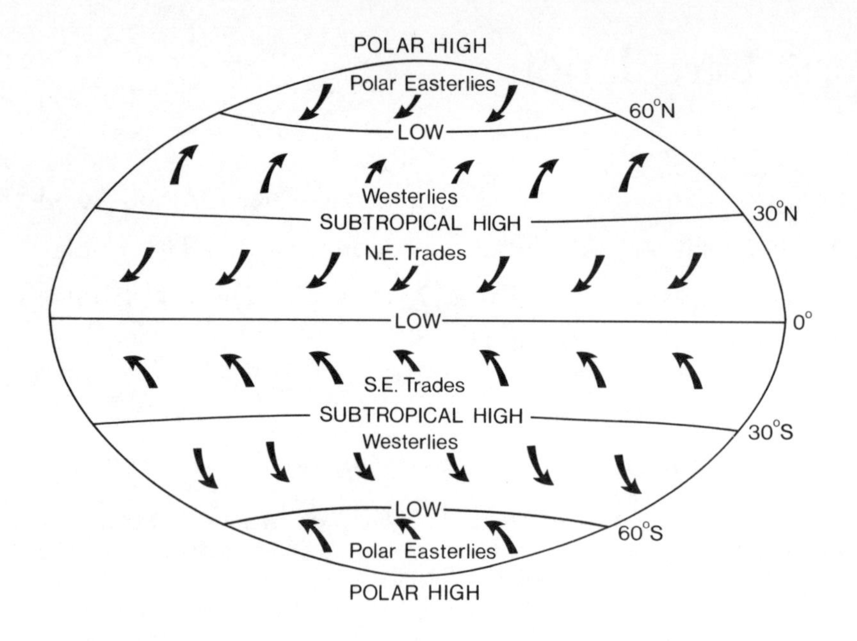

12. 3 Schematic representation of zonal pressure belts and wind systems near the Earth's surface.

of the mid-latitudes. One is by a system of large horizontal motions in vortices, and the other is by the development of waves in a zonal flow (Fig. 12. 4). In each case the part of the flow with a component towards the pole carries warm air and that with a component towards the equator carries colder air, and thermal exchanges take place with adjacent air at the poleward and equatorward limits of the flow.

DISHPAN EXPERIMENTS AND THE HADLEY CELL

The simple convection cell illustrated in Figure 12. 1 can be simulated in a cylindrical vessel, or 'dishpan', containing a shallow layer of water which is heated at the periphery of the base (which represents the equator), and cooled at the centre of the base (which represents the pole). If the dishpan is then rotated about a vertical axis, various more complicated motions (relative to the pan) can be produced. These depend upon the temperature gradient in the liquid between the rim and the centre of the pan, and on the speed at which the pan is rotated. As the rotation rate is increased the flow becomes zonal (Fig. 12. 5-a), and then at higher rotation speeds it follows a series of large loops with closed circulations between them (Fig. 12. 5-b).

On the spherical Earth the component of the Earth's rotation about the local *vertical* axis is least in the lowest latitudes; like the CorF it is proportional to the sine of the latitude. Hence it is in the lowest latitudes that a thermal convection cell may be expected, and this in fact is the case where the so-called *Hadley Cell* exists (Fig. 12. 6). Pressure is low along the thermal equator in the lower atmosphere, and air converges here and rises. In the sub-tropics air descends and there is high pressure near the surface, giving an equatorward pressure gradient. The winds which this causes are deflected to the west by the CorF, giving the fairly steady north-east and south-east trade winds in the northern and southern hemispheres respectively.

THE UPPER WESTERLIES, JET STREAMS AND ROSSBY WAVES

In middle and high latitudes the winds in the upper troposphere are essentially westerly and exhibit patterns similar to those shown in Fig. 12. 5-a and -b. These upper westerlies are almost geostrophic or gradient winds. As temperature generally decreases polewards throughout the troposphere the thermal wind is from the west in both hemispheres, and thus the speed of

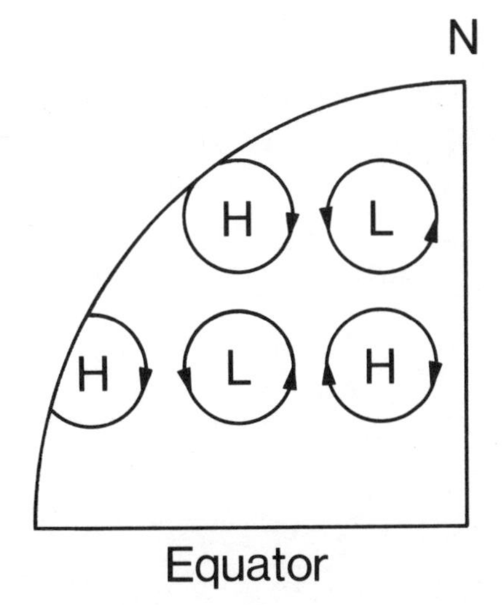

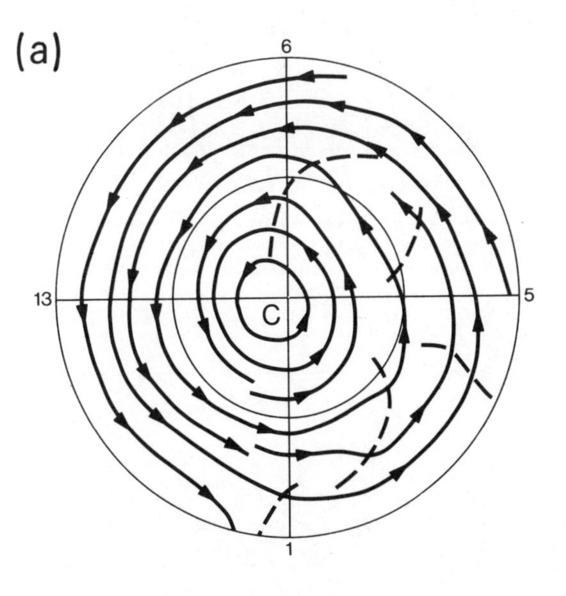

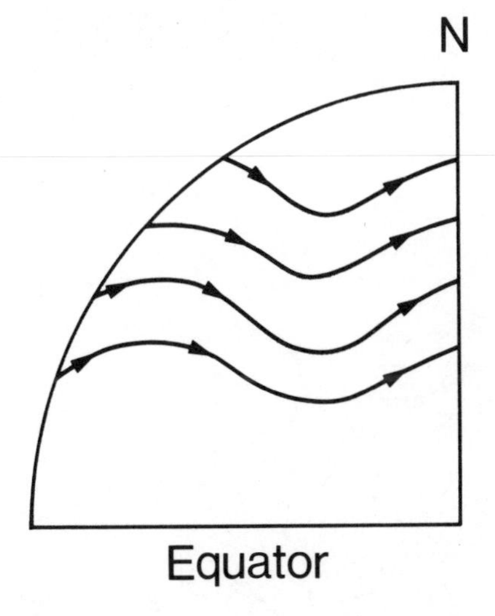

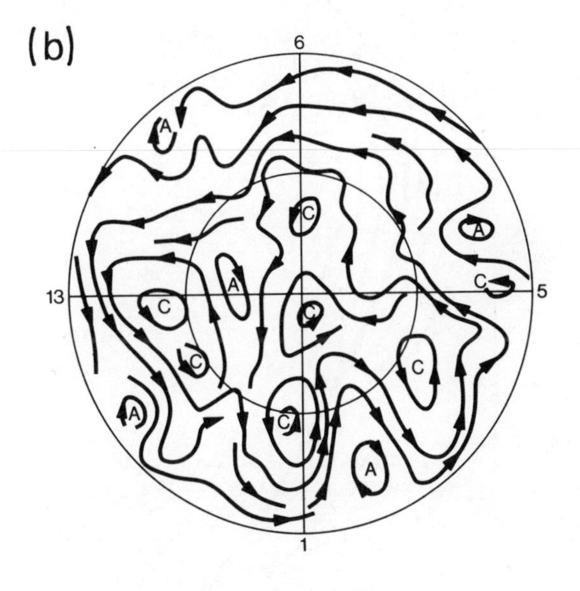

12. 4 Possible mechanisms bringing about a poleward heat transfer: (top) motion in vortices, (bottom) zonal waves. (After S. Petterssen)

12. 5 Relative streamlines illustrating flow in two dishpan experiments: (a) transition between Hadley and Rossby regimes, (b) steady-state Rossby regime. (After D. Fultz et al.)

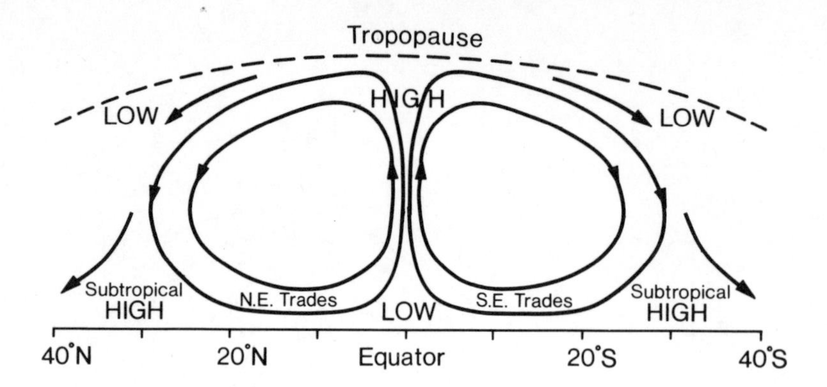

12.6 Schematic representation of the Hadley Cells either side of the equator.

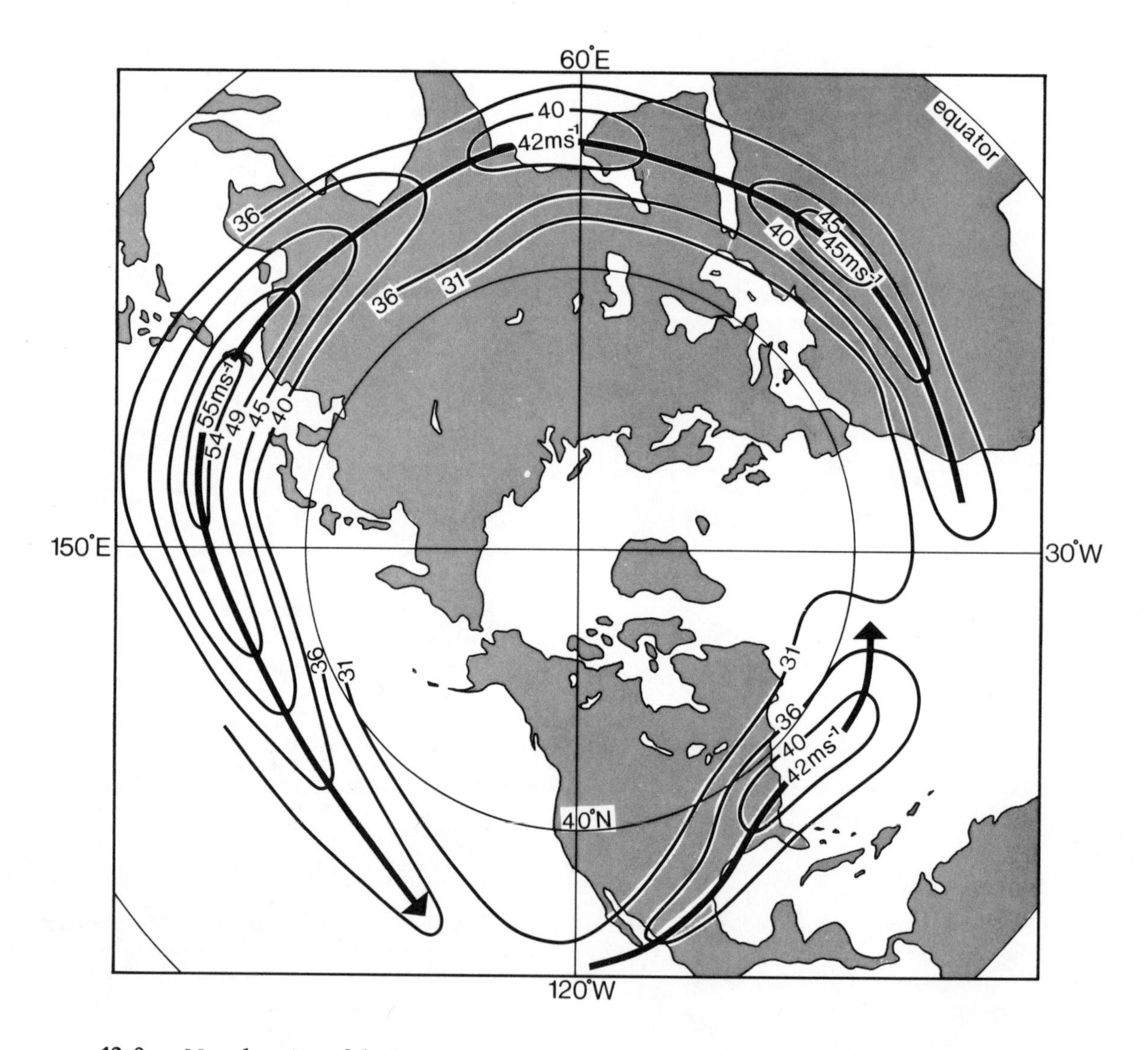

12.8-a Mean location of the jet stream (geostrophic wind speeds in m s-1 at the level of maximum speed) in January.

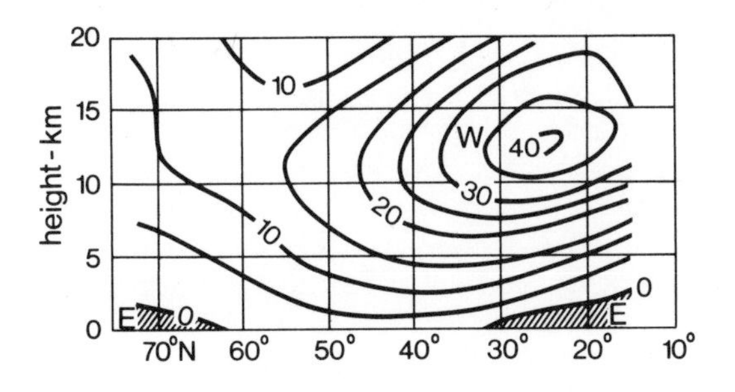

12. 7-a Zonal wind speed (m s^{-1}) in winter, averaged around the world.

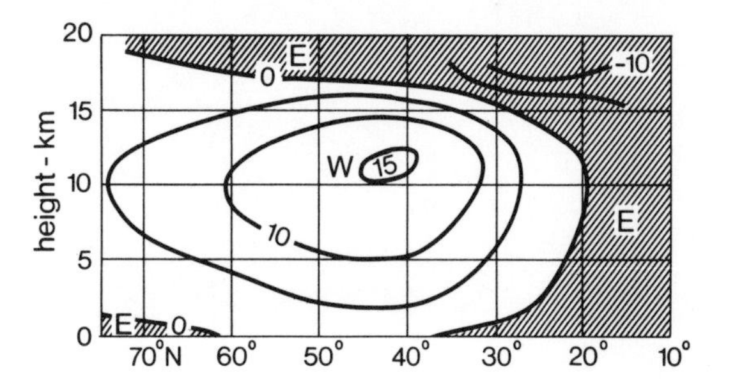

12. 7-b Zonal wind speed (m s^{-1}) in summer, averaged around the world. (After S. Petterssen)

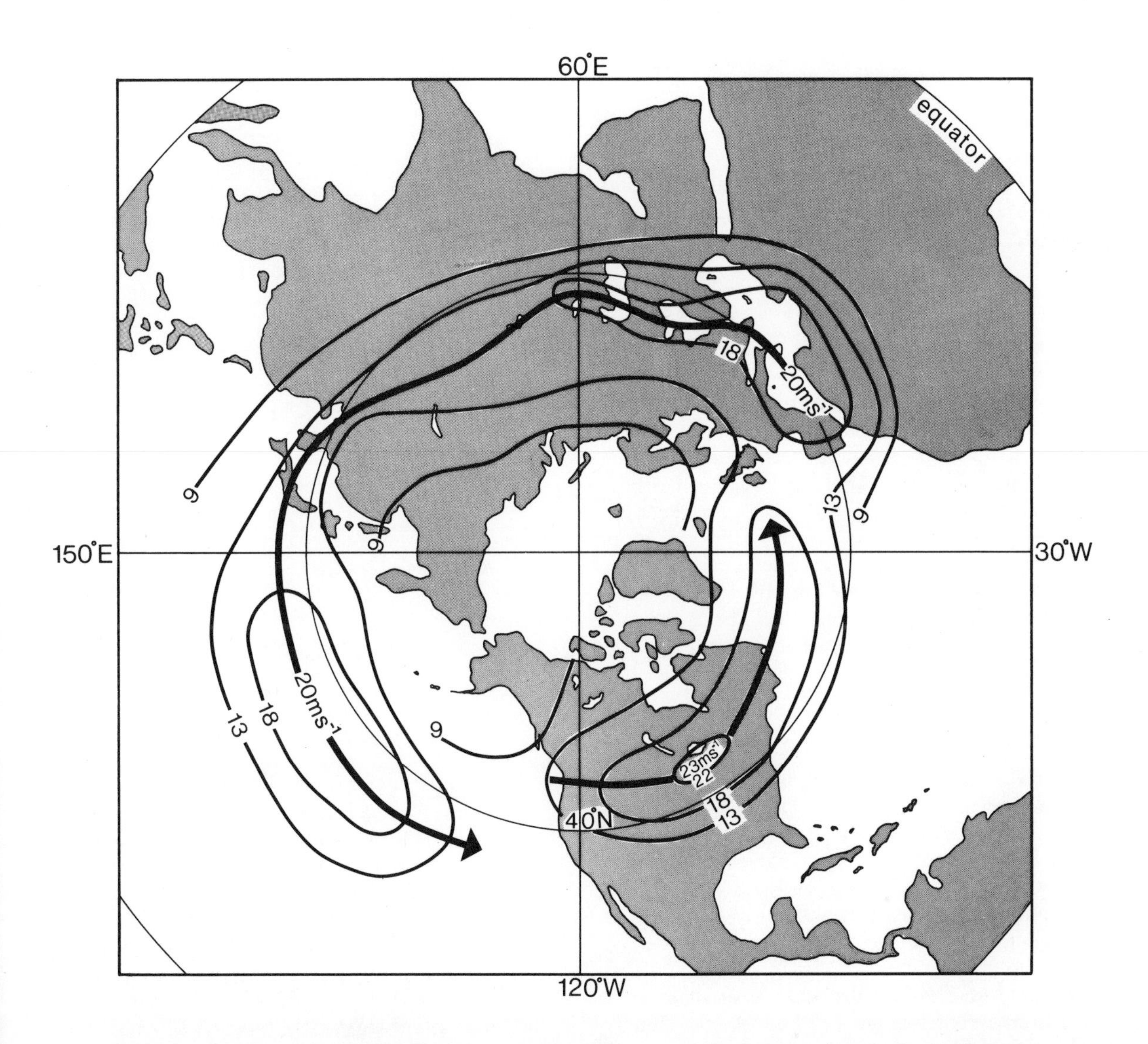

12. 8-b Mean location of the jet stream (geostrophic wind speeds in m s^{-1} at the level of maximum speed) in July. (After Namias and Clapp)

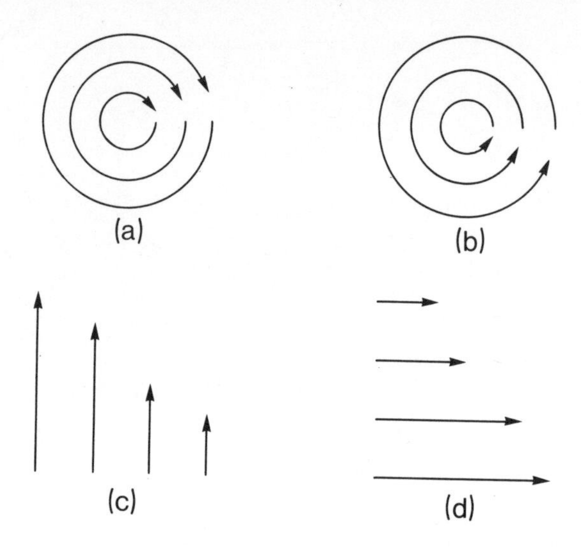

12. 9 Examples of streamlines with negative vorticity, (a) and (c), and positive vorticity, (b) and (d).

the westerly wind increases with altitude. The thermal wind is strongest in regions where the poleward temperature gradient is greatest. In the vicinity of fronts between air masses, where horizontal temperature gradients may reach as much as 1°C per 10 km at a particular level, it becomes extremely strong with speeds sometimes exceeding 100 m s^{-1} (approximately 195 knots) in winter. Such strong winds exist in narrow belts just below the tropopause and are known as *jet streams*. Their presence may be detected from the drift of cirrus clouds, and their main features have become apparent since aircraft started flying at these high altitudes in World War II. The average zonal wind speed in the troposphere (Fig. 12. 7) shows that the jet stream is centred at between 10 and 15 km height, and maps of the mean location of the jet stream (Fig. 12. 8) indicate winds of up to 50 m s^{-1} circling the globe in middle latitudes.

The actual pattern on any one occasion is, however, likely to depart considerably from these mean situations. Two separate jet streams may often be distinguished, the *polar front jet* and the *subtropical jet*, and perhaps others associated with additional frontal zones (e.g. an Arctic front). The individual jets are much narrower and stronger than Figure 12. 8 would suggest, but are often discontinuous and follow somewhat tortuous paths resembling the streamlines in Figure 12. 5-b. The meanders which occur in streamlines in dish-pan experiments and in the upper westerlies in the atmosphere are known as *Rossby waves*. The parts where they approach closest to the pole are called *ridges*, and those where they are nearest the equator are *troughs*.

To understand why Rossby waves form in the upper westerlies we must introduce the concept of *vorticity*. By vorticity is meant a tendency to spin about an axis. Here we shall be concerned with vorticity relative to a vertical axis, and by convention we shall take anti-clockwise rotational tendency viewed from above as positive vorticity and clockwise rotational tendency as negative vorticity (Fig. 12. 9).

Fluid moving on the Earth's surface possesses vorticity about a vertical axis as a result of the Earth's rotation if it is anywhere other than at the equator. This is the *planetary vorticity*, and its magnitude is equal to the Coriolis parameter, f. It is positive in the northern hemisphere and negative in the southern hemisphere. The fluid may also possess vorticity as a result of its motion relative to the Earth – a cyclone in the northern hemisphere possesses positive vorticity. This is the *relative vorticity* and is denoted by the symbol ζ. The sum of f and ζ is known as the *absolute vorticity*.

In large-scale horizontal motion in which vertical motion may be ignored, the conservation of angular momentum leads in the absence of friction to the absolute vorticity remaining constant,

i.e. $$\frac{d}{dt}(f+\zeta) = 0 .$$

Thus if a westerly wind is diverted polewards in the northern hemisphere, f increases and ζ will tend to decrease, and the air will gain anticyclonic vorticity. This will turn the air equatorwards, but f will then decrease so that the air gains cyclonic vorticity causing it to swing back polewards. The wind thus swings backwards and forwards about its mean latitude in a series of long waves. Rossby was able to show that these waves may be *stationary* (i.e. their troughs and crests remain in the same position) when there is a particular relationship between their wavelength and the wind speed. This is important, because the waves are usually initiated by topographic features (e.g. the Rockies), and if they remain fixed relative to those features they can become persistent factors bringing anomalously warm conditions in a ridge and cold conditions in a trough. For the wind speeds typically experienced in the upper westerlies, the wavelengths for stationary Rossby waves are such that there are usually three or four waves in the circumpolar westerly flow in middle latitudes, with the number increasing as the mean latitude of the flow decreases and the length of the path round the Earth increases.

THE MID-LATITUDE WESTERLIES

We may now look further at the surface winds in mid-latitudes and consider the causes of the subtropical high pressure and the sub-polar latitude low pressure with which they are associated. An important feature of these winds, particularly in the northern hemisphere, is their variability. As we noted in Chapter 7, winds from all directions may be experienced in the British Isles, and it may be only when we take observations over an appreciable length of time that we find the predominance of westerlies or south-westerlies. The reason is that the pressure distribution in these latitudes varies considerably. In particular, the mean low pressure centres in latitudes of about 60° in Figure 12. 2 result from the generally eastward movement of a succession of individual depressions whose paths may be shifted somewhat to the north or south, and which may at times be replaced by high pressure cells.

These individual and travelling pressure systems usually originate as a result of divergence or convergence in the upper westerlies. As the upper westerlies pass through a ridge in a Rossby wave they experience anticyclonic curvature and their speed, given by the gradient wind speed equation (10. III. b), is greater than the geostrophic wind speed. But when they pass through a trough, they experience cyclonic curvature and their gradient wind speed is less than that of the geostrophic wind. There are other factors which will also bring about accelerations and decelerations in the upper westerlies, but in general we find that their speed is increasing as we proceed from a trough to a crest, and vice versa.

Where speed increases in the direction of flow, divergence must occur. This causes air to be drawn up from below, creating a low pressure area or depression at the surface. Conversely,

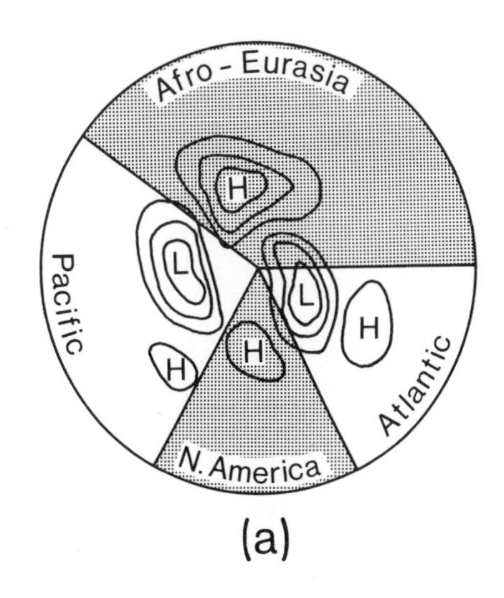

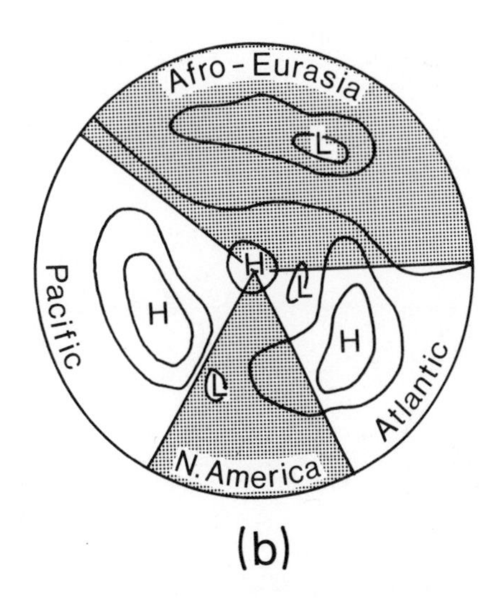

12. 10 Idealised pressure distribution in the northern hemisphere, shown on a four-sector model: (a) winter, and (b) summer.

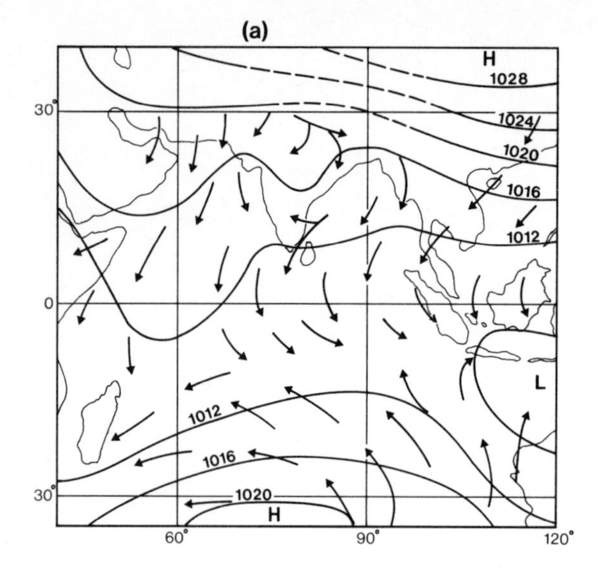

12. 11-a Surface winds and pressure distribution reduced to sea level over Southern Asia and the Indian Ocean in January.

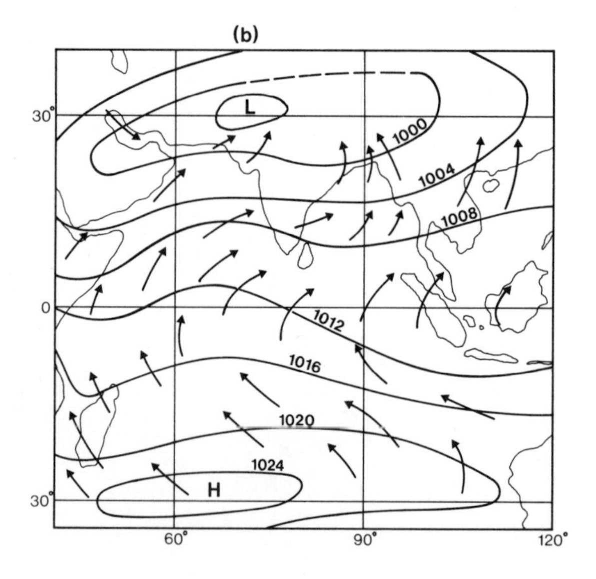

12. 11-b Surface winds and pressure distribution reduced to sea level over Southern Asia and the Indian Ocean in July. (After H. Riehl)

where speed decreases in the direction of flow, convergence occurs and air sinks to form an anticyclone with high pressure at the surface. The subsequent movement of these pressure centres is determined essentially by the upper westerlies, and the depressions therefore move eastwards and polewards towards the ridges in the Rossby waves while the anticyclones move eastwards and equatorwards to merge with the subtropical high. We shall look further at the weather conditions associated with these depressions and anticyclones in Chapter 13.

INFLUENCE OF CONTINENTS; MONSOONS

The presence of continents can disturb the features shown in Figure 12. 3 through either thermal or dynamic causes. The dynamic effects are at their greatest when the continents have mountain ranges. A westerly air stream flowing over a mountain range across its path is deviated polewards whilst ascending but equatorwards while descending, and thus a wave trough forms downwind of it. The presence of the Rockies is therefore partly responsible for the development of depressions over the Atlantic and for the low pressure area around Iceland which is shown in Figure 12. 2.

A more obvious way in which the presence of continents affects the atmospheric circulation pattern is through the thermal contrasts which exist between land and sea and which were discussed in Chapter 5 (see Fig. 5. 11). In summer the air over the land is warmer and less dense than that over the ocean, and thus low pressure areas tend to be centred over continents while high pressure centres are mainly located over the oceans. In winter, when the air is cooler and denser over the land than over the ocean, the situation is reversed. The contrast between summer and winter is particularly noticeable in the northern hemisphere and can be represented by a simplified model with four quadrants, two of water and two of land (Fig. 12. 10).

Associated with such seasonally alternating pressure systems are seasonally reversing winds known as *monsoon winds*. In principle their causes are essentially the same as those of land and sea breezes, but they exist on much larger time and distance scales and interact with the winds of the upper troposphere. The regions of the Earth in which they are found may be identified from comparison of Figures 12. 2-a and -b. The most notable region is that of the Indian subcontinent and the adjacent Indian Ocean, and this is the example which will be considered here; but other monsoon winds are found along the West African coast between latitudes 5° and 15°N, the China Sea, and northern Australia. The development of monsoon winds is closely related to changing patterns in the upper winds and to displacements of jet streams, but we will confine ourselves here to the surface winds, pressure distribution and the weather associated with the Indian monsoon.

In winter, surface pressure is high over the continent of Asia. This gives a pressure gradient from north to south over India and the Indian Ocean to about 10°S (Fig. 12. 11-a). North of the equator the winds are predominantly north-easterly – the direction of the trade winds in the Hadley cell. When they cross the equator they are deflected to the left by the CorF, and back to a north-westerly direction. These winds coming from the continental mass to the north bring dry and cool weather.

In spring the land mass begins to warm, and pressure falls over northern India. As temperatures rise convection produces violent storms, but at this stage there is little moisture available and it is generally dry. It is not until late May or June that the south-westerly monsoon 'bursts' with south-westerly winds bringing humid maritime air. This has been drawn across the equator from the southern Indian Ocean by the PGF caused by the very low pressure over the Indo-Gangetic Plain (Fig. 12. 11-b), acquiring moisture over the Arabian Sea on its way to India. The burst of the monsoon is accompanied by heavy rain associated with orographic influences along the western side of India, and convective storms in the interior. Monsoon depressions move westwards across the northern part of India bringing precipitation to the Ganges valley.

13.0 Twin hurricanes Ione and Kirsten moving across the Pacific Ocean between Hawaii and Mexico. The central 'eyes' of the storms are well-defined in this photograph taken by a satellite in a polar orbit in August 1974.

13 Cyclones and Anticyclones

Mid-latitude frontal depressions □ Formation in relation to upper westerlies □ Life history and associated weather □ Tropical cyclones □ Formation, warning signs, associated weather □ Avoiding action for shipping □ Anticyclones: formation and associated weather.

AS WE HAVE SEEN in Chapter 12, large eddies, both cyclonic and anticyclonic, are essential features of the general circulation pattern of the atmosphere. They may be considered to be responsible for the intermediate zones of high and low pressure between the equator and the poles which appear on maps of mean sea level pressure, or they may be considered to result from them. In particular, the formation of mid-latitude cyclones, or *depressions* as they are often called, is closely associated with the frontal zone produced by the convergence of tropical and polar air masses, a feature of the general circulation of the atmosphere. In this chapter we shall consider the major aspects of mid-latitude depressions and anticyclones and also of tropical cyclones, and the weather characteristic of them.

MID-LATITUDE DEPRESSIONS

Before upper-air observations indicating the presence of jet streams were available, a model for the development of depressions in relation to the polar front had been constructed by the Bergen school of meteorologists in Norway. The chief features of this model are shown in Figure 13. 1. The cyclone or depression forms where a wave develops in the polar front which allows a tongue of warmer tropical air to penetrate into the polar air mass. The whole system is travelling eastwards, and thus a warm front along which warm air displaces colder air can be distinguished from a cold front following it at which the reverse occurs. The air masses are converging along the polar front, and the warm air tends to rise over the cold air at the warm front while a wedge of cold air undercuts the warm air at the cold front. The ascent of air creates low pressure at the surface with isobars surrounding the centre of low pressure. The winds near ground level blow across the isobars at an angle determined by the nature of the surface, as discussed in Chapter 10, and thus air spirals inwards and upwards towards the centre of the depression. As the air in the warm sector is gradually lifted, the cold front catches up with the warm front and the depression becomes occluded. There is still, warm air aloft, and the isobars and winds show a cyclonic pattern, but the only frontal contrast at the surface is between the fresh cold air at the rear of the depression and the modified cold air in front. Such an occlusion can be either cold or warm, depending on the way in which the cold air in the front of the depression has been modified. These depressions depend upon the conversion of potential energy to kinetic energy for their continuance, and dissipate when the frontal contrast between adjacent air masses disappears. Being associated with the ascent of air and thus with condensation, and also with strong winds, depressions have a considerable influence on weather conditions (Fig. 13. 2). These can mainly be considered as features of the warm and cold fronts of the depression.

Warm fronts (Fig. 13. 3) typically have slopes of less than 1 in 100, and thus the ascent of air at a warm front is gradual (unless the warm air becomes unstable), and stratiform clouds will form. The first sign of an approaching warm front is the appearance of cirrus clouds at a height of perhaps 10 km. As the front continues to approach, the clouds become lower and thicker. The frontal surface itself acts as a stable layer, sometimes as an inversion, and thus any clouds forming in the cold air below the front will be of limited vertical

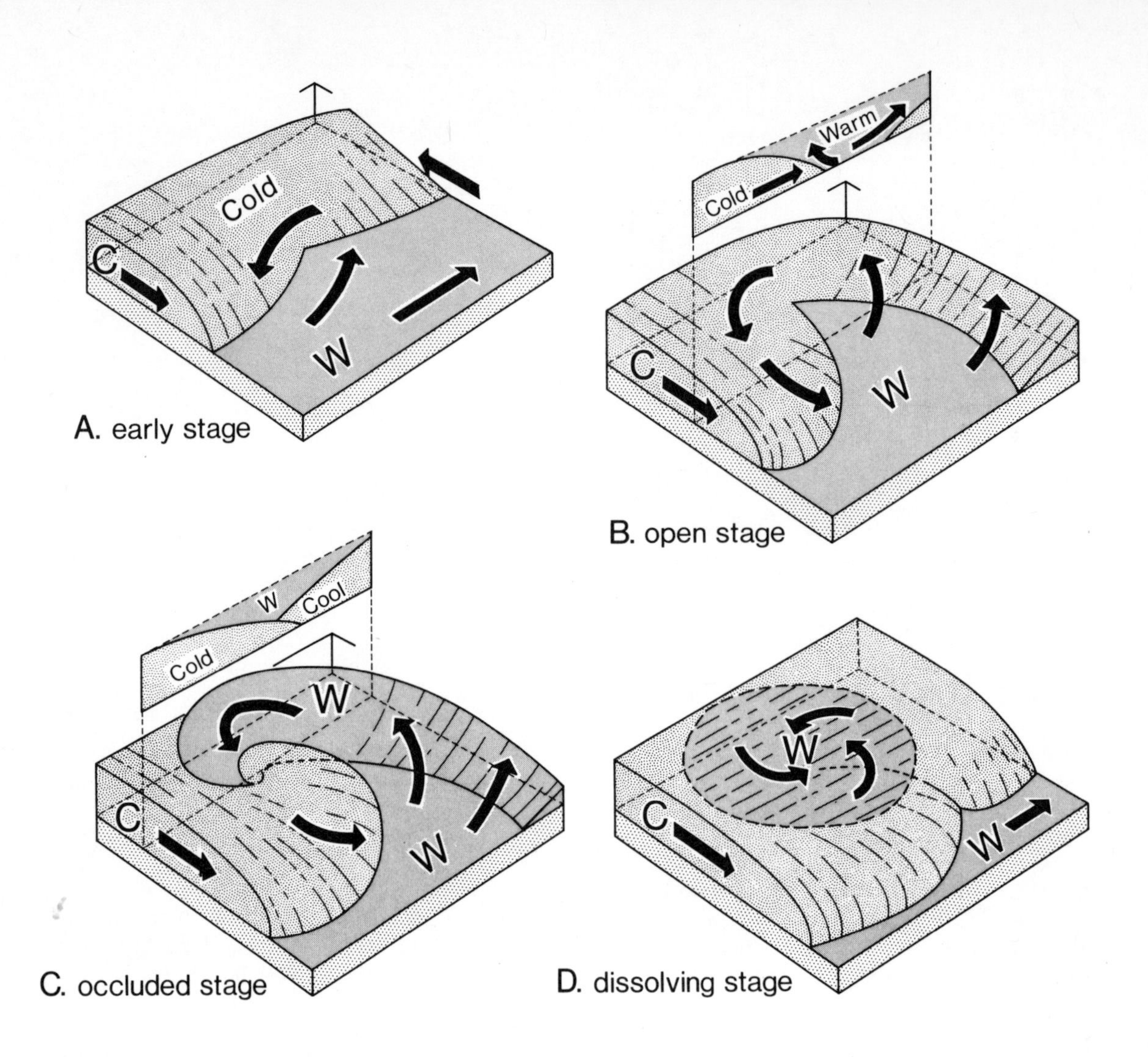

13.1 The life cycle of a middle-latitude cyclone according to the Bergen model. (After A. N. Strahler)

extent. Altostratus clouds will perhaps be associated with drizzle some 500 km ahead of the front at the surface, giving way to nimbostratus with heavier rain as the front gets nearer. Ahead of the warm front pressure falls steadily, temperature rises slowly, and the winds in a northern hemisphere depression will chiefly blow from the south in increasing strength, but at the passage of the front they will veer to south-west.

In the warm sector of the depression the temperature, the pressure and the wind remain fairly steady. The amount of cloud and precipitation depend very much on the characteristics of the warm air mass. It will usually be stable, but (particularly if it is forced to rise by an orographic barrier) stratus cloud and drizzle may be experienced, and on a sunny afternoon cumulus clouds may develop. Over a relatively cold sea surface advection fog is likely to occur, and in any case visibility is usually poor.

When the *cold front* (Fig. 13.3) approaches there is a marked change in the weather. Cold air is pushing under warm air, and the slope of the cold front is considerably greater than that of the warm front – perhaps 1 in 50, and greater near the surface where the front's progress is being

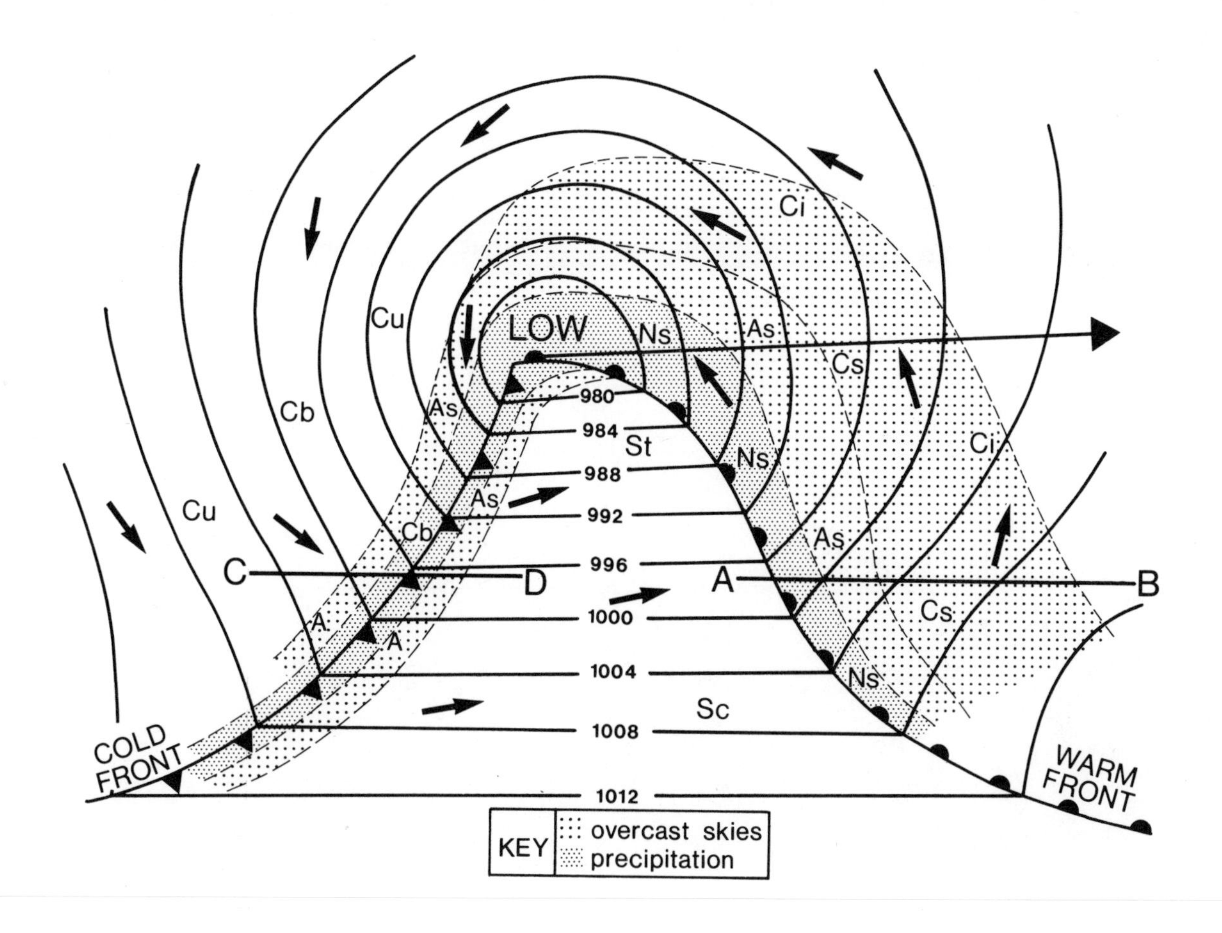

13. 2 A typical frontal depression in the northern hemisphere, showing isobars (pressure in mb), fronts, winds, clouds and areas of precipitation.

retarded by f riction. Upcurrents are thus more violent, and instability is much more likely to develop, with towering cumulonimbus clouds and heavy precipitation in the form of showers, perhaps with thunder. The wind typically veers sharply to north-west or north in a northern hemisphere depression at the passage of the front, and can become very strong in squalls. The temperature falls sharply, while the pressure begins to rise, and visibility improves markedly except in the showers. The weather behind the cold front will depend on the characteristics of the cold air mass. The mass will usually be unstable, giving bright but showery weather.

Occlusions (Fig. 13. 4) are also accompanied by cloud and precipitation. A warm occlusion exhibits similar features to a warm front, but superimposed somewhat ahead of the surface front is an upper cold front with cumulus or cumulonimbus clouds and showers. At the passage of a cold occlusion some of the aspects of a cold front are experienced, but these are preceded (and to some extent followed) by the clouds characteristic of a warm front; the clearance of clouds after the passage of a cold occlusion is much slower than in the case of a cold front.

Southern hemisphere frontal depressions may be considered to be essentially the same as their counterparts in the northern hemisphere, but whereas the wind always *veers* at fronts in the northern hemisphere, it *backs* in the southern hemisphere, from north to north-west at a warm front and from west to south-west at a cold front.

Such models allow weather forecasts for frontal depressions to be made for periods of up to 24 hours ahead if synoptic observations are available

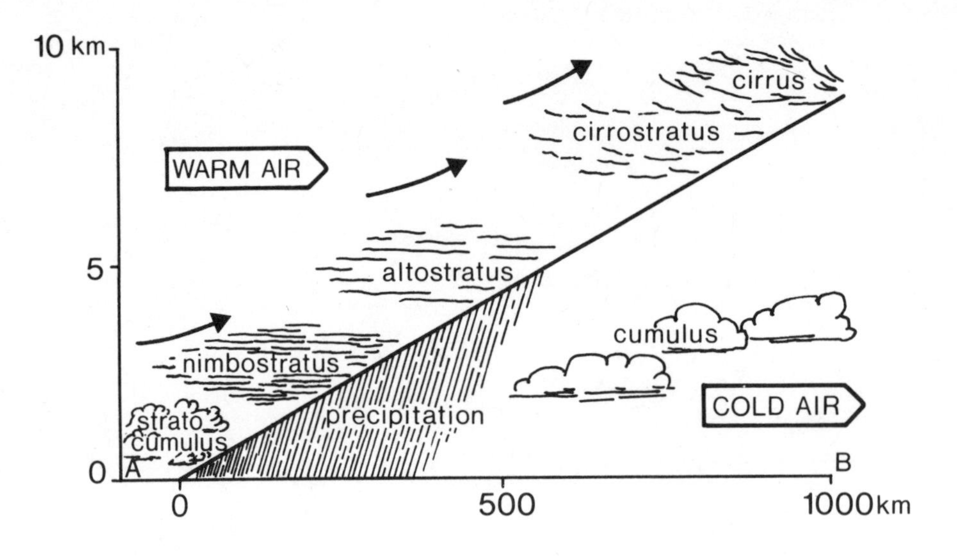

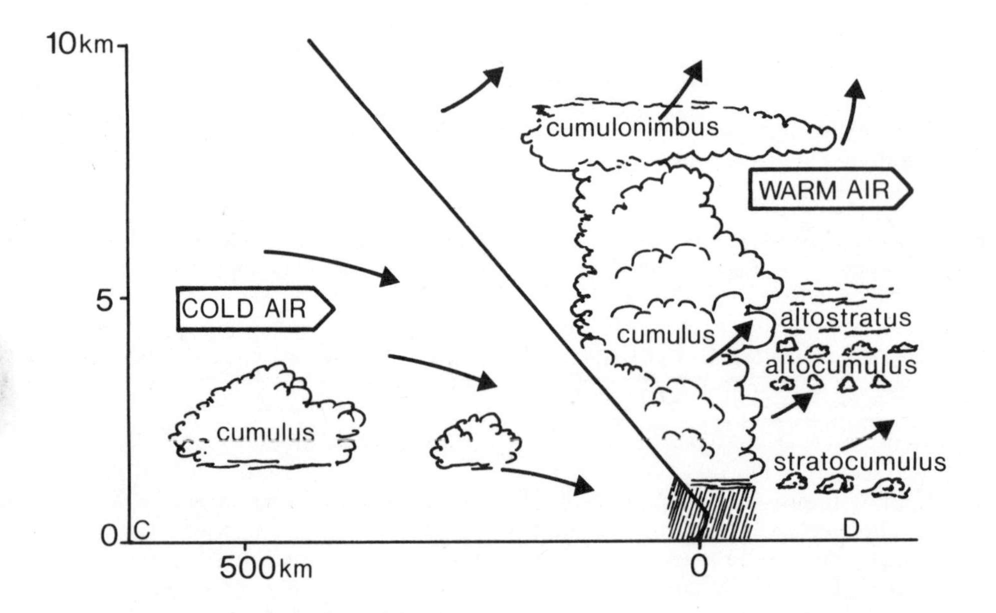

13. 3 Vertical sections through (top) a warm front, and (bottom) a cold front of a mid-latitude depression. Positions of sections (AB and CD) are shown on Figure 13. 2.

for sufficiently wide areas. For Europe such observations must extend to the west over the adjacent Atlantic Ocean because these depressions typically travel 1 000 km per day – a rough guide to the movement of frontal depressions is given by the direction and 70% of the speed of the surface geostrophic wind in the warm sector. Each individual depression has its own unique characteristics, and no idealised model can adequately represent a particular depression, but the main features of the model described above can usually be recognised. The availability of upper air data has assisted considerably in making forecasts for individual frontal depressions. These rely on knowledge of the relationship between divergence in the upper westerlies and depression formation (or *cyclogenesis* as it is called) and its prediction from vorticity considerations as discussed in Chapter 12, and of the rôle of the upper westerlies in steering depressions.

In addition to the frontal depressions described above, various other types can be recognised in mid-latitudes. Frontal depressions frequently travel in families along the line of the polar front, and often a secondary depression becomes involved in the circulation pattern of a larger depression adjacent to it, and moves around it in a cyclonic direction. The commonest type of secondary depression is that which has formed behind the primary depression on the trailing edge of its cold front, and has caught up with it. Usually each successive member of a family of depressions follows a course which is on the equatorward side of its predecessor. Mid-latitude non-frontal depressions include *lee depressions* which form in the lee of major topographical barriers; *thermal lows* which result from unequal heating particularly of land areas in summer; and *polar air depressions* which sometimes develop in cold unstable air passing over a warm sea surface.

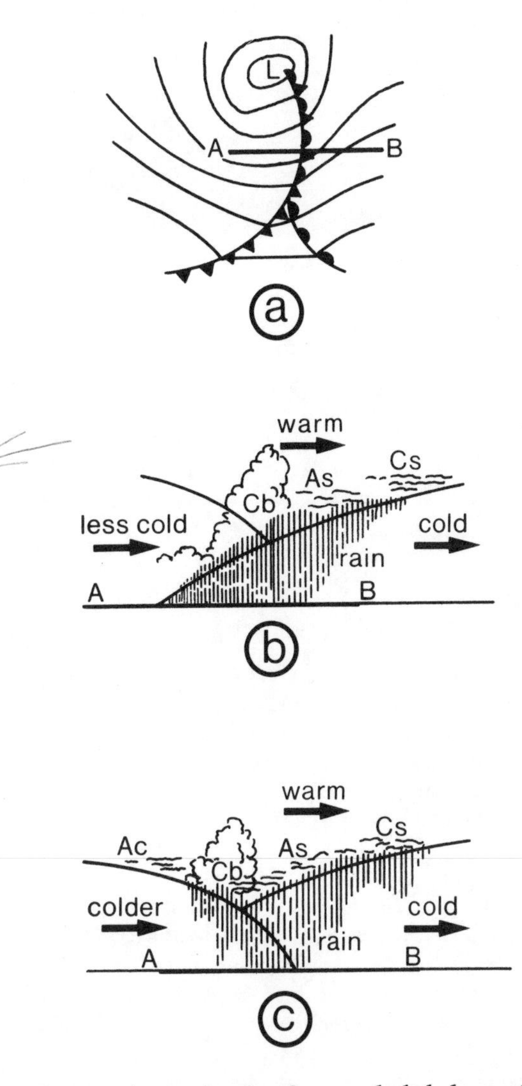

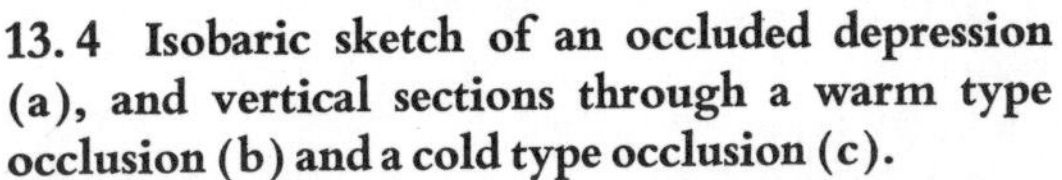

13.4 Isobaric sketch of an occluded depression (a), and vertical sections through a warm type occlusion (b) and a cold type occlusion (c).

TROPICAL CYCLONES

These are relatively small but very intense features, known also as *hurricanes*, *typhoons*, and by other local names in particular parts of the world, and are generally referred to by mariners as 'tropical revolving storms'. By definition they have winds reaching Beaufort Force 12 (33 m s^{-1} or 64 knots and over). Less intense features are referred to as 'tropical depressions' or 'tropical storms'. These cyclones do not have a frontal structure, but their winds are usually very much stronger than those of mid-latitude depressions, often exceeding 50 m s^{-1} (more than 100 knots), so that they are a serious hazard to shipping and cause considerable damage if they travel over land areas. Their diameters are normally less than half those of typical mid-latitude depressions, but the pressure at their centre is likely to be 960 mb or less, some 50 mb below that at their periphery, compared to a difference of perhaps 30 mb between the pressure at the centre and periphery of a mid-latitude depression.

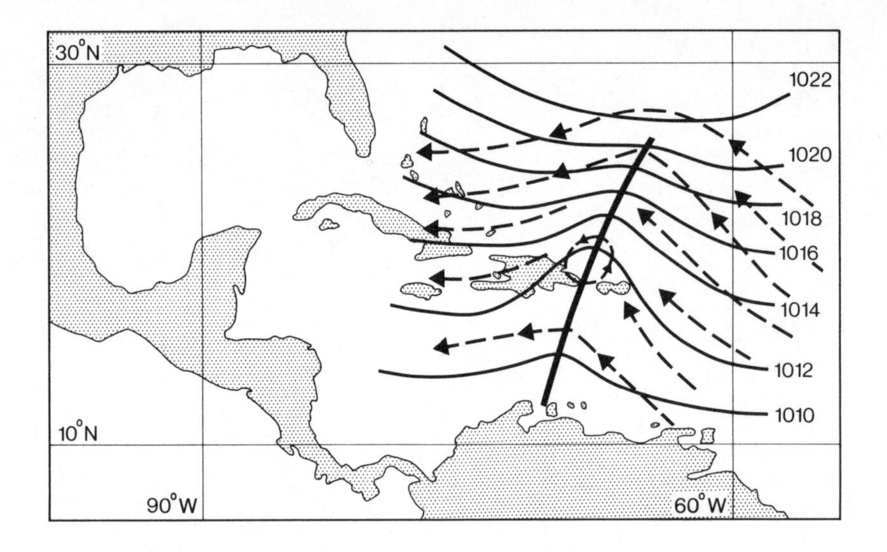

13.5 Model of a wave in the easterlies showing pressure in mb at the surface (continuous lines) and streamlines at about 4 600 m (broken lines). (After H. Riehl)

As there are no marked frontal contrasts in low latitudes, tropical cyclones must develop from dynamic causes coupled with thermal instability. The first stage in their formation is usually the development of a wave in the easterly air flow of the trade winds over the ocean. This is associated with a trough of low pressure extending into the trade wind zone from the Equatorial Trough (Fig. 13.5). In front of the trough line there is horizontal convergence, air rises, and condensation forms clouds. Many such easterly waves form, but few develop into tropical cyclones. A number of conditions have been identified as being essential for their development. These include an extensive ocean surface with temperature greater than 27°C, a deep layer of moist and unstable air, a latitude of at least 5°, and a small vertical wind shear. The first and second of these conditions are required to supply the storm with sufficient potential energy, the third condition is necessary for the cyclonic circulation to be established – CorF becomes zero at the equator – and the fourth condition is necessary for the storm to develop vertically without distortion. Even when all these conditions are satisfied only a small proportion of initial disturbances develop into full tropical cyclones – perhaps a total of 50 per year over all of the oceans. The other vital ingredient appears to be the presence of an anticyclone in the upper troposphere, providing divergence above and thus promoting inflow below.

The tropical cyclone, once formed, is in essence radially symmetrical (Fig. 13.6). The violent winds spiral inwards, counterclockwise in the northern hemisphere and clockwise in the southern hemisphere, but at the centre there is an area of light winds and generally clear sky which is known as the 'eye' of the storm. Furthest from the centre of the cyclone, spiral bands of cirrus and cirrostratus clouds are encountered. Moving towards the centre, the clouds become thicker and there is heavy precipitation. The eye of the storm is surrounded by dense and massive cumulonimbus clouds, and it is here that air ascends rapidly, water vapour condenses to release enormous quantities of latent heat, and there is torrential rain and thunderstorms. The fully developed tropical cyclone has a warm core due to this release of latent heat, and this intensifies the anticyclone in the upper troposphere which is essential for the convergence at low levels to continue. In the eye itself, which typically has a diameter of perhaps 40 km, there is some subsidence of air, and this leads to adiabatic warming.

The paths of tropical cyclones are largely controlled by the subtropical high pressure areas over the oceans. Initially they move westwards, but then often curve polewards around the high pressure centre and finally travel in an easterly direction. Their rates of movement are slower than those of mid-latitude depressions. At first they generally travel at less than 15 knots, but after

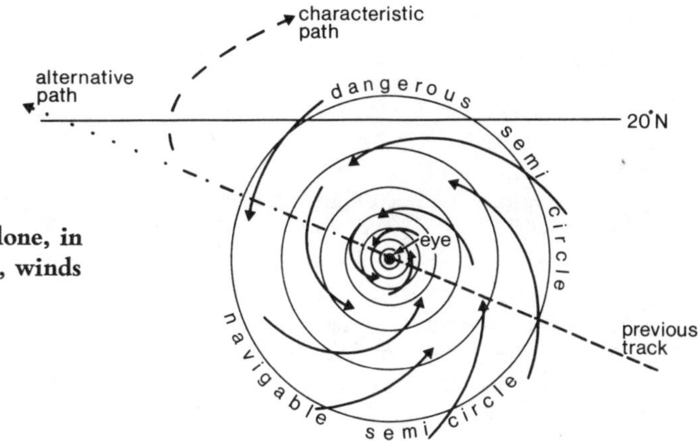

13. 6 A tropical revolving storm, or cyclone, in the northern hemisphere, showing isobars, winds and possible paths.

recurving eastwards their speed may increase to 25 knots or more. Their decay is almost always attributable to their moving away from their source of heat and moisture. If they encounter a land area before recurving eastwards they can bring enormous destruction, but are usually short-lived because their supply of moisture is cut off and frictional drag at the surface is increased. After recurvature over the ocean, they move over a progressively cooler sea surface and decay more gradually. Typical paths of tropical cyclones are shown in Figure 13. 7. It will be seen that they do not affect the South Atlantic or eastern South Pacific Oceans. This is explained by the relatively low surface temperatures, and by the related fact that even in the southern summer the Equatorial Trough does not lie more than 5° south of the equator in these parts of the ocean. The seasons for tropical cyclones are generally the three or four months following the summer solstice, when ocean surface temperatures are highest, but in the Arabian Sea and Bay of Bengal they may occur in almost any month except February and March.

The winds of tropical cyclones generate very large waves which move outwards from the storm centre. As the storm advances more waves are being generated, and the sea becomes extremely confused, particularly behind the eye of the storm. Mariners speak of a *navigable semi-circle,* which is

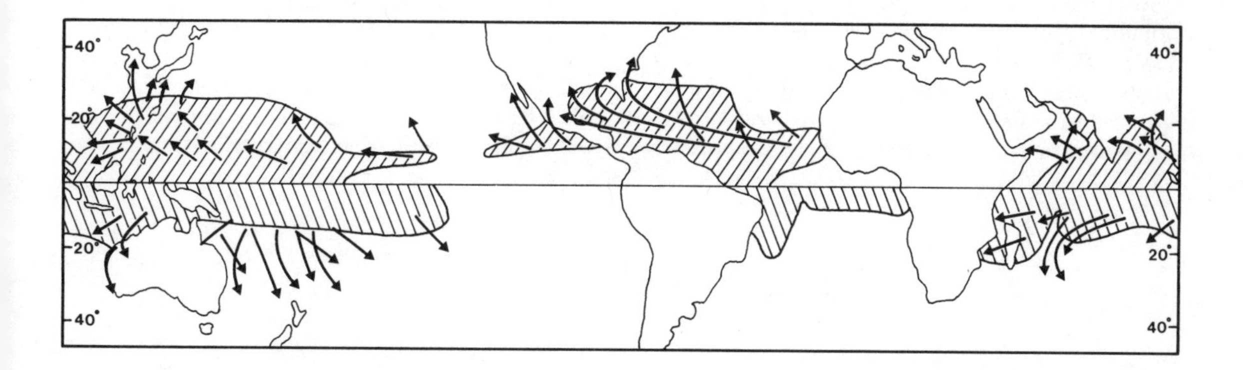

13. 7 Characteristic tracks of tropical cyclones. The shaded areas are those ocean areas with mean sea-level temperature exceeding 27°C in September (northern hemisphere) or March (southern hemisphere). (After T. Bergeron)

on the equatorward side of the cyclone's track while it is travelling westwards, and contrast it with the *dangerous semi-circle* in which a ship will be blown towards the cyclone's path and in which the effective wind fetch is greatest because the winds are blowing in the same direction as that in which the cyclone is travelling. There exist detailed instructions to enable mariners to recognize the approach of a tropical cyclone, and suggesting what action they should take to avoid the dangerous semi-circle and the centre of the storm if they find themselves in its vicinity. (See *Meteorology for Mariners*, Met. Office 593, and Admiralty Pilots.)

ANTICYCLONES

Being associated with subsiding air and divergence, anticyclones are the source regions of air masses, and are generally characterised by dry, quiet weather. Even though the gradient wind speed is greater for an anticyclone than for a cyclone with the same horizontal pressure gradient (Chap. 10), anticyclones usually have light winds because their pressure gradients are relatively small. At the surface the wind, which blows clockwise in the northern hemisphere and anticlockwise in the southern hemisphere, blows outwards across the isobars. Air thus diverges, air mass properties are almost constant, and fronts are unlikely to be encountered. The subsiding air warms adiabatically, and thus its relative humidity decreases and any clouds present tend to evaporate. Towards the bottom of the subsiding air there is likely to be a temperature inversion, and this restricts the vertical development of any cloud which might form by turbulent mixing in the lowest layers or by convection on a sunny afternoon. Anticyclones are commonly slow-moving or stationary, and so can bring settled weather for a few days or weeks at a time. They do provide conditions in which such features as land and sea breezes in summer and radiation fog in winter can develop.

Large-scale anticyclones exist over the subtropical oceans, and over continental areas in high latitudes in winter. Anticyclones, sometimes of a less permanent nature, are often encountered in mid-latitudes in summer and in winter. They can be classified as either cold or warm. A cold anticyclone owes its high pressure to the air in the lower layers of the atmosphere being colder than the surrounding air, and can thus be quite a shallow feature. A warm anticyclone must result from convergence at high levels in the atmosphere providing a greater mass of air above it than above the surrounding region, and is thus essentially dynamic in origin and related to the air flow in the upper troposphere. The subtropical belts of high pressure and the anticyclones formed beneath the Rossby waves in the upper westerlies are of this type.

Cold anticyclones occur mainly over land areas and over the polar regions, and normally have a strong temperature inversion. This serves to trap dust and pollutants in the lowest layers of the air, giving poor visibility and an anticyclonic gloom, which readily becomes smog when combined with radiation fog. When the air remains clear, severe frosts are common. Warm anticyclones over land in summer are associated with clear sunny weather, and the combined effects of solar radiation and adiabatic warming can produce very high temperatures. Over the sea the surface temperatures are less subject to variation, and there is less contrast between the weather conditions associated with anticyclones in summer and winter. Clouds resulting from vertical mixing by turbulence are often encountered with anticyclones in mid-latitudes, particularly in winter when the sky may be overcast, but they are of limited vertical extent, and precipitation from them is rare.

13. 8 The Earth from 35 800 km above Brazil. Portions of four continents are visible in this ATS-3 satellite photograph. South America dominates the lower half of the picture, with North America at upper left, and Europe and Africa at upper right. A cold front can be seen moving across the United States, and a cylonic depression is located above Northern Europe. There are large areas of clear sky in the sub-tropical regions over the Atlantic Ocean, where surface pressure is high and air is descending.

14.0 A dye tank at Woods Hole Oceanographic Institute. It is used to study the effects of winds, solar heat and the Earth's rotation on oceanic circulation. The tank rotates about a vertical axis while vents and lamps provide winds and heat, and coloured ink is released into the water to show up the patterns of flow. Land masses and sea-bed topography can be simulated by rubber barriers. Such experimental models are used frequently in order to examine ways in which the motion of water masses is affected.

14 Oceanic Circulation

Frictional coupling between the atmosphere and ocean □ Ekman spiral and storm surges □ Circulation of upper layers of the ocean □ Movement of icebergs □ Convergence and divergence □ Upwelling □ The deep circulation.

MOST CURRENTS IN THE OCEAN other than tidal currents derive from either the stress of the wind on the water surface or the uneven distribution of mass due to variations in temperature and salinity. In general, the currents in the upper layers are attributable to wind stress, whereas those below are part of the thermohaline circulation resulting from the distribution of temperature and salinity, but this distinction is often unclear, and the two are connected by vertical water movements.

WIND-INDUCED CURRENTS, THE EKMAN SPIRAL AND STORM SURGES

The wind blowing across a water surface exerts a frictional stress in the direction in which it is blowing as well as generating waves. There is some forward movement of water particles in wave motion, and the magnitude of the stress which the wind exerts is very dependent on the roughness of the water surface, so the two processes are closely interrelated.

The simplest situation in which to consider the nature of the currents resulting from wind stress is that in which a wind blows at constant velocity over a deep, infinite and homogeneous ocean, so that the water movement is not impeded by any continental barriers and the coefficient of (eddy) viscosity remains constant with depth. We could reasonably assume that in such a situation the speed of the resulting current would be greatest at the surface and would decrease with increasing depth, and that below some depth at which the current became negligibly small we could ignore friction. Considering the layer above this depth as a whole, and assuming that the wind remains steady for a period long enough to allow a steady current to become established with no acceleration, we can equate the magnitude of the wind stress (τ) with the CorF and conclude that the mean current ($\bar{u}$), and therefore the net transport of water in this layer, is 90° *cum sole* from the wind direction (Fig. 14. 1). Thus

$$\tau = D\rho f\bar{u} \ ,$$

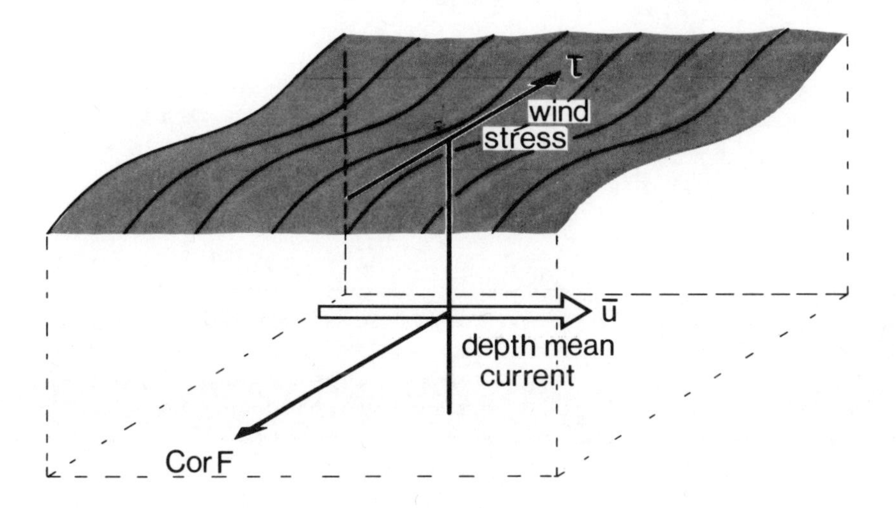

14. 1 The balance of forces in the Ekman layer of the ocean to give a depth mean current 90° to the right of the direction of the wind stress in the northern hemisphere.

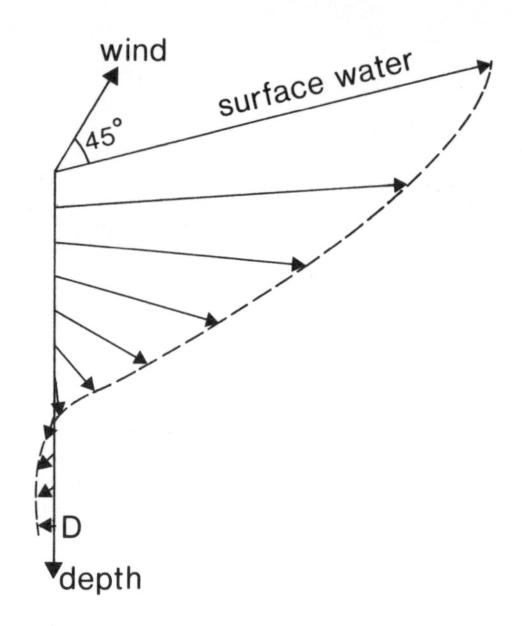

14.2 The Ekman spiral, showing variation of current velocity with depth in the Ekman layer.

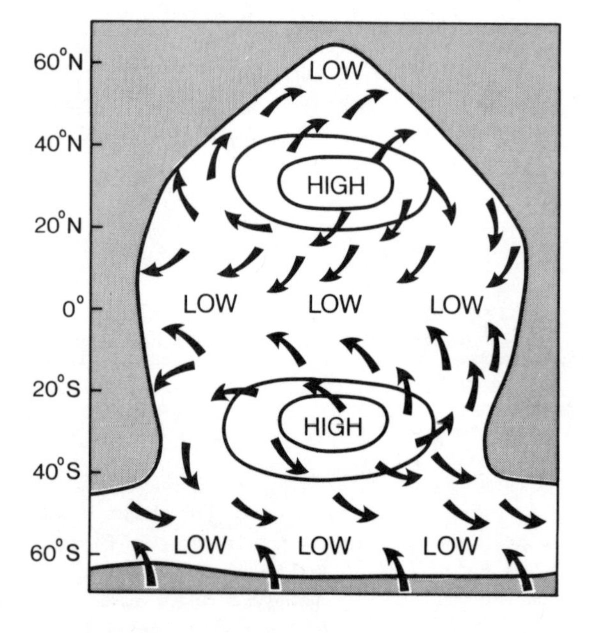

14.3 A diagrammatic representation of the pressure distribution and winds over an idealised ocean.

where D is the depth of the layer so that $D\rho$ is the mass of the layer per unit surface area. Hence

$$\bar{u} = \frac{\tau}{D\rho f} \, . \qquad (14.1)$$

At any one depth, however, the water is subject to three forces – the stress of the overlying water (or, at the surface, the wind), the stress of the underlying water, and the CorF. The effect of this is that the current deviates increasingly *cum sole* with increasing depth. Ekman, who examined this point theoretically, found that the surface current deviates 45° *cum sole* from the wind direction, and that the speed of the current decreases exponentially with depth as well as turning further *cum sole* until, at the depth D, it has about 4% of the surface speed and is directed exactly opposite to the surface current. This current structure is known as the *Ekman Spiral* (Fig. 14.2), and the depth D, which varies with the eddy viscosity and the latitude, is called the *depth of frictional influence* and is typically between 100 m and 200 m. The speed of the surface current depends on the same factors as D, and on the wind stress, and is usually between 1% and 3% of the wind speed.

The complete structure of the Ekman Spiral has never been observed in the oceans, partly no doubt because the assumptions on which it is based (a steady wind, an infinite and homogeneous ocean, and no other forces acting) are somewhat unrealistic. Observations of the surface current well away from the land have, however, shown speeds similar to those predicted by Ekman and deviations *cum sole* of the wind direction, though usually less than 45°. The classic example is the drift of the ship *Fram* in the Arctic ice which was 20° to 40° to the right of the wind, but this information was available to Ekman when he studied the problem and his theory was intended to provide an explanation of it.

Other atmospheric parameters can affect water movement as well as wind. The sea surface appears to react as an inverted barometer to atmospheric pressure variations. An atmospheric low is thus accompanied by high sea level, and if the winds associated with the low cause water to move towards a coast against which it piles up, particularly high water levels can be experienced. Both mid-latitude depressions and tropical cyclones are moving systems; if their speed is appropriate, such high water levels or *storm surges* can travel with them as long waves. This occurs fairly often in the North Sea where a storm surge, accompanying a depression moving rapidly east and with its centre passing just to the north of Britain, travels round the coast of Scotland and then follows an anti-clockwise path round the edge of the North Sea. Its effects are greatest where the North Sea narrows and shoals in the south, and by the time the storm surge reaches here the northerly winds in the cold sector of the depression may well be causing further water to flow southwards in the North Sea. It was such a situation, combined with spring tides, that led to disastrous flooding on the east coast of England and in Holland on 31 January and 1 February 1953.

THE SURFACE CIRCULATION OF THE OCEANS

The Atlantic and Pacific Oceans are essentially similar in shape, extending northwards from the Southern Ocean and narrowing at their northern ends, and can be represented diagrammatically as in Figure 14. 3. The winds between about 10° and 50° latitude are essentially anticyclonic around the subtropical high. This leads to water in the Ekman layer above the depth of frictional influence being transported towards the centre of the ocean. The convergence here depresses the main thermocline, and is associated with features described in Chapters 6 and 7, but it also leads to the water level sloping downwards from the centre of the ocean outwards, and this gives rise to a gradient current which is anticyclonic, i.e. running in the same direction as the wind (Fig. 14. 4).

This *subtropical gyre* is very asymmetrical, particularly in the northern oceans, its centre being displaced to the western side of the ocean, so that the Gulf Stream in the North Atlantic and the Kuroshio in the North Pacific are very much stronger than any currents on the eastern side of these oceans. The reason for this involves vorticity considerations which were discussed in Chapter 12. If we simplify the situation appreciably by ignoring vertical motion, we can consider the factors which bring about changes in vorticity as the water moves round the subtropical gyre. Throughout the gyre the water is acquiring negative vorticity in the northern hemisphere from the anticyclonic wind stress at the surface. The water on the eastern side is moving equatorwards, and its wind-induced negative vorticity may be just sufficient to fit it to the lower positive planetary vorticity at lower latitudes, so that its absolute vorticity is conserved. On the western side, water is moving polewards, so that its vorticity becomes negative relative to the Earth and its absolute vorticity becomes increasingly negative. Some braking action is needed to prevent vorticity increasing indefinitely, and this can only be supplied by friction, either at the lateral boundaries or at the sea floor, or by viscosity within the water. This requires much higher velocities on the western side, as friction is proportional to something like the square of the water speed. The result is that the speeds are typically some ten times greater in the warm western boundary current than in the cool eastern boundary current, and the western boundary current extends to greater depths; but it is less than one twentieth

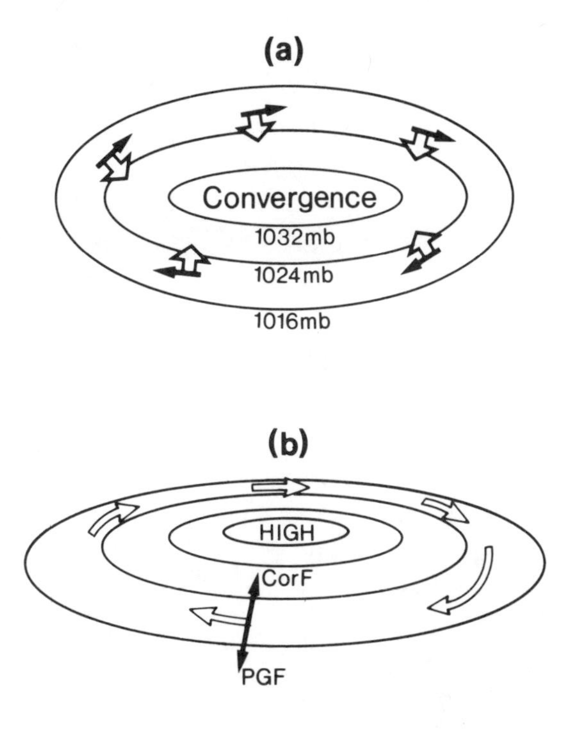

14. 4 Water movements associated with anticyclonic winds in the northern hemisphere: (a) atmospheric pressure and winds, and associated Ekman transports to the right, (b) resultant topography of the sea surface, and associated gradient currents below the depth of frictional influence.

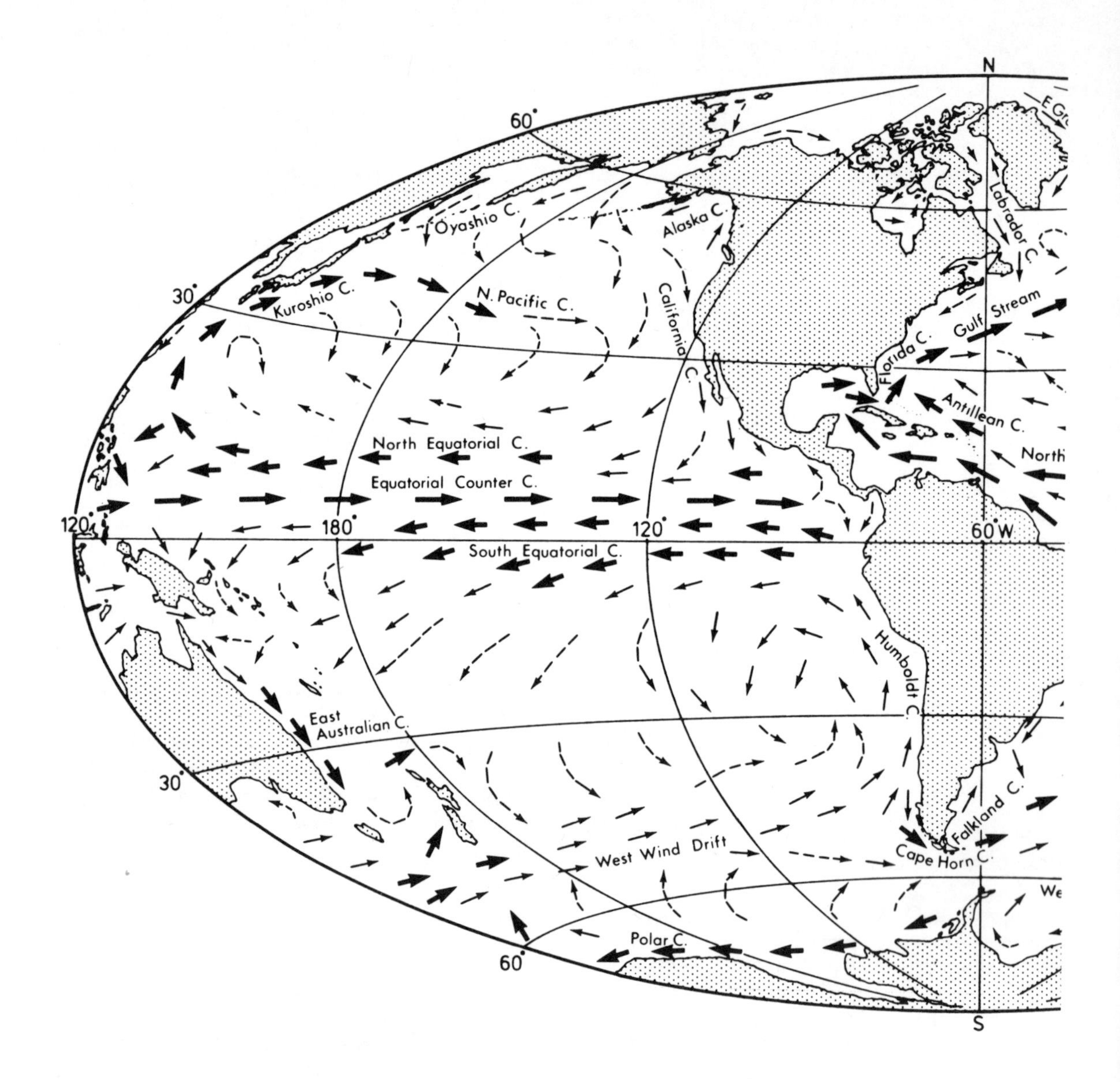

14. 5-a The currents at the surface of the world ocean in the northern winter (left half).

the width of the ocean and water flows equatorwards in most of the remainder, so that the same amounts of water are being transported in both directions. The asymmetry of the subtropical gyres south of the equator is less well marked, and in the South Pacific the cold Peru current on the eastern side is perhaps the dominant feature. The different pattern of distribution of land and sea in the southern hemisphere appears to be responsible for this contrast.

In the trade wind zone, water is transported across the oceans to the western side, and thus there is a slope of the sea surface from west to east. Along the inter-tropical convergence zone, where

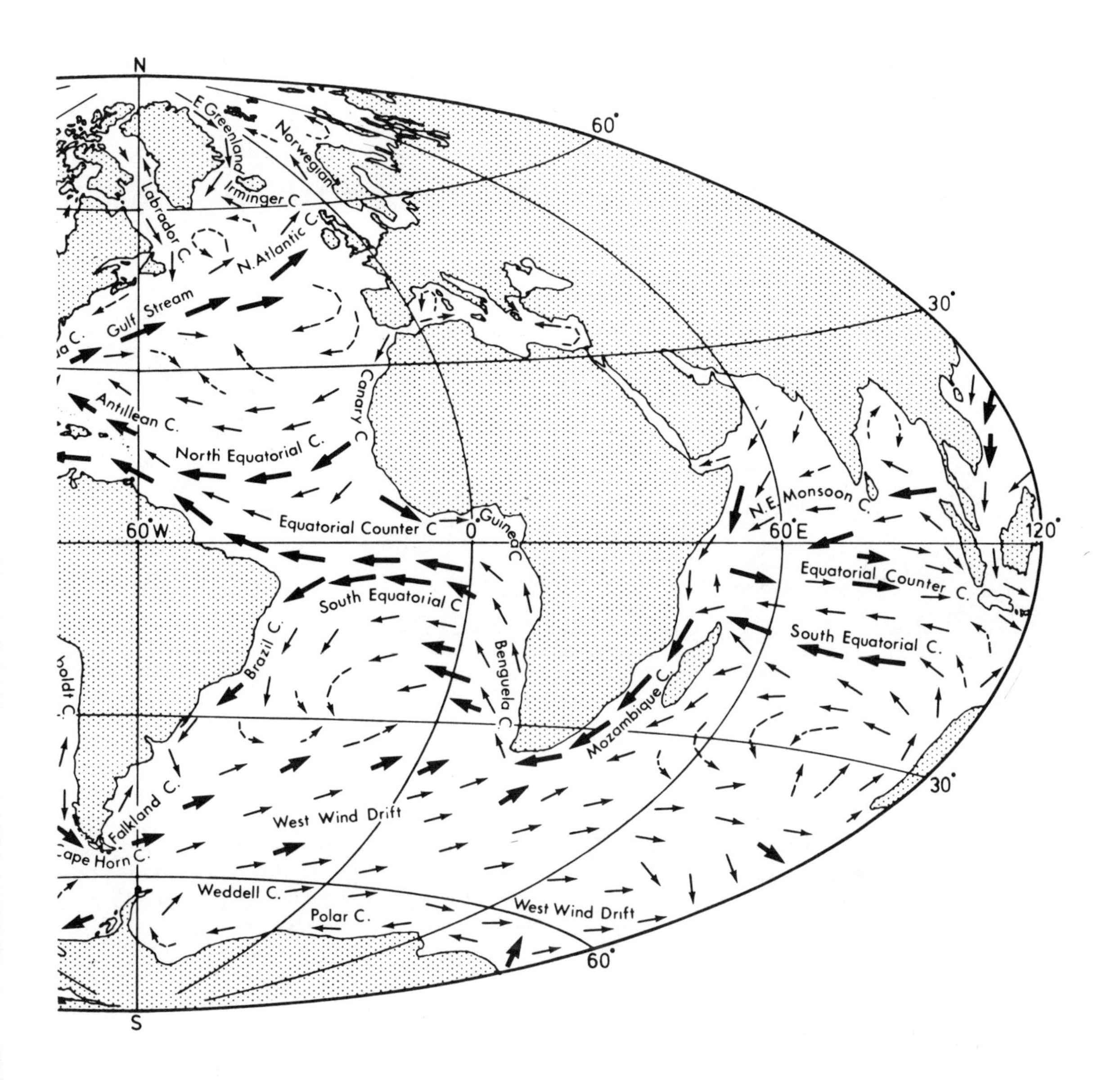

14. 5-a The currents at the surface of the world ocean in the northern winter (right half).

winds are light, water is able to flow back downhill to the east, and so near to the equator the CorF has little effect on it. This current is known in each ocean as the *Equatorial Counter Current*. At the equator itself, where the CorF changes direction, there is divergence at the water surface. This brings the thermocline nearer to the surface and also promotes vertical mixing in it, which leads to the water above the thermocline being *denser* at the equator than on either side, but that below the thermocline being *less dense* at the equator than on either side. This causes the pressure gradients at the depth of the thermocline (about 100 m) to slope away from the equator to

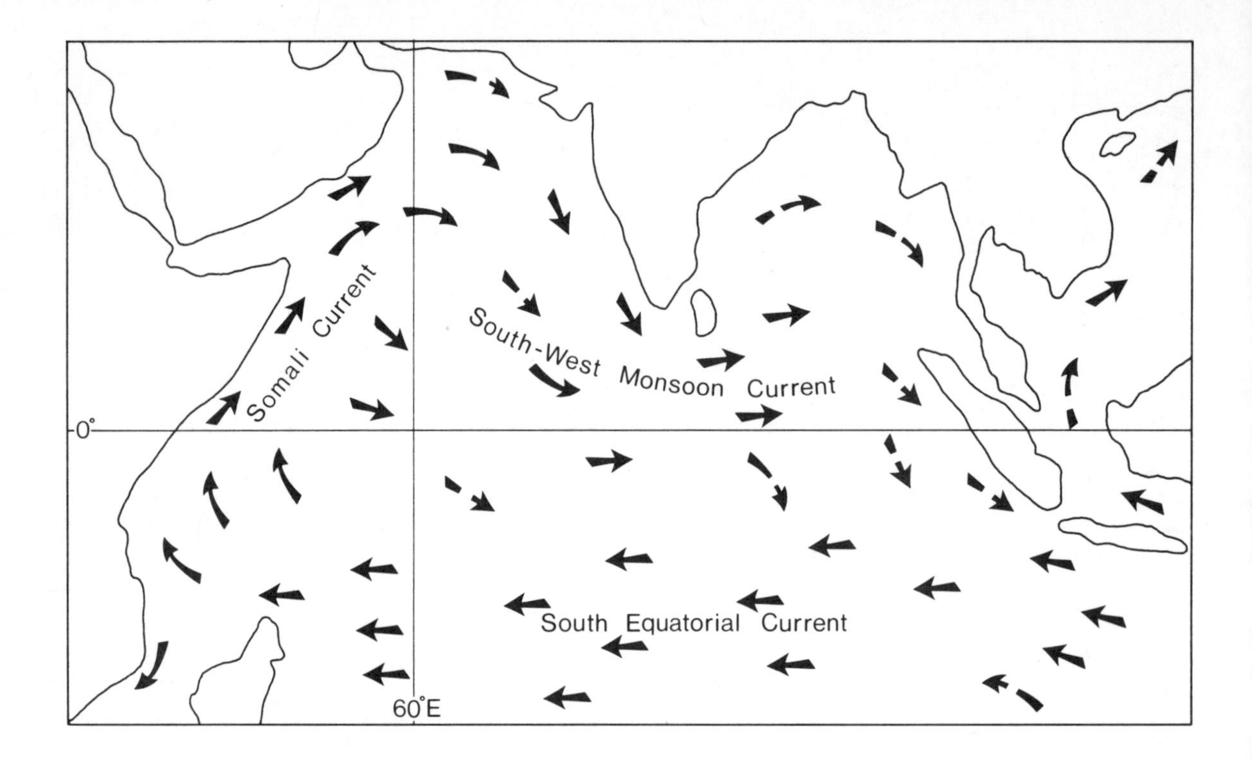

14. 5-b The currents at the surface of the Indian Ocean in the northern summer. (After G. Dietrich)

either side, which in turn causes geostrophic currents towards the east, both immediately north and south of the equator. Further, on either side of the equator, the directions of the wind-induced currents at a depth of 100 m will be nearly opposite those of the surface currents (Ekman spiral effect), and although very weak they will lead to convergence at the equator and will contribute to the flow there with an easterly component. There is also the pressure gradient towards the east resulting from the sloping water surface, and all these factors, together with the absence of any CorF actually at the equator, appear to be responsible for the *Equatorial Undercurrents* which have been observed in each of the oceans. These currents are comparable to the Gulf Stream in terms of their speeds and the volumes of water which they transport, but it was not until the 1950s that they were discovered and first investigated, and their dynamics are still incompletely understood.

At latitudes greater than 50° there is a marked contrast between the northern and southern hemispheres. In the North Atlantic and North Pacific, water movement is obstructed by continental barriers, but in the southern hemisphere it is able to travel right round the globe in the *Antarctic Circumpolar Current*. Over the Southern Ocean the winds are essentially westerly, giving net water transport away from the Antarctic continent. This causes the sea surface to slope upwards towards the equator, and the associated gradient currents run towards the east, again in the same direction as the wind is blowing. There is, however, a very important convergence zone surrounding Antarctica lying generally somewhere between 50°S and 60°S and known as the *Antarctic Convergence*. There is a marked increase in sea-surface temperature as one proceeds northwards across this Convergence, and it is here that water sinks to form *Antarctic Intermediate Water*. Its location has been related to the zone of strongest winds over the Southern Ocean, but its position is too constant for it to be attributable entirely to the wind field, and there are various theories relating it to the circulation of sub-surface water.

The icebergs which originate in the Weddell and Ross Seas may initially move westwards in currents close to the Antarctic continent associated with easterly winds of the high latitudes, but will then be carried eastwards in the major circumpolar current. They reach furthest north in November and December, when they may be encountered at perhaps 40°S in the Falkland Current of the South Atlantic.

In the North Pacific there is a *subpolar gyre* comprising the Alaska and Oyashio Currents which is comparable to the subtropical gyre to the south of it, but in the opposite sense associated with the cyclonic wind stress in latitudes 50°–70°. In the North Atlantic the equivalent subpolar gyre is interrupted by Greenland and exists in two parts, one on either side of Greenland. The East Greenland Current carries pack ice and icebergs from the glaciers which reach the east coast of Greenland southwards. Off Cape Farewell this current converges with the warm Irminger Current, and most of the ice soon melts, though some may persist in the cool inshore part of the northerly-flowing West Greenland Current. To this is added a considerable input of icebergs from glaciers which reach the West Greenland coast, particularly in North-East Bay and Disko Bay. These continue in the subpolar gyre of the Labrador Sea and Baffin Bay, and in due course travel southwards in the Labrador Current to reach the Grand Banks off Newfoundland. The greatest numbers are encountered here in spring and early summer, having been released perhaps two years earlier from a fjord in the spring thaw and then spent their last winter frozen in ice between Baffin Land and Labrador. Off Newfoundland the larger icebergs will ground, and gradually they will break up and melt in this region of convergence between the warm Gulf Stream and the cold Labrador Current.

The Indian Ocean extends less far to the north than the Atlantic and Pacific Oceans, and is subject to the major seasonal wind reversal of the monsoons. The effect of the monsoons on the surface circulation of the Indian Ocean can be seen by comparison of Figure 14. 5-a and -b.

VERTICAL WATER MOVEMENT AND THE DEEP CIRCULATION

Zones of convergence and divergence in the flow of the upper layers of the ocean have already been noted. Divergence occurs particularly

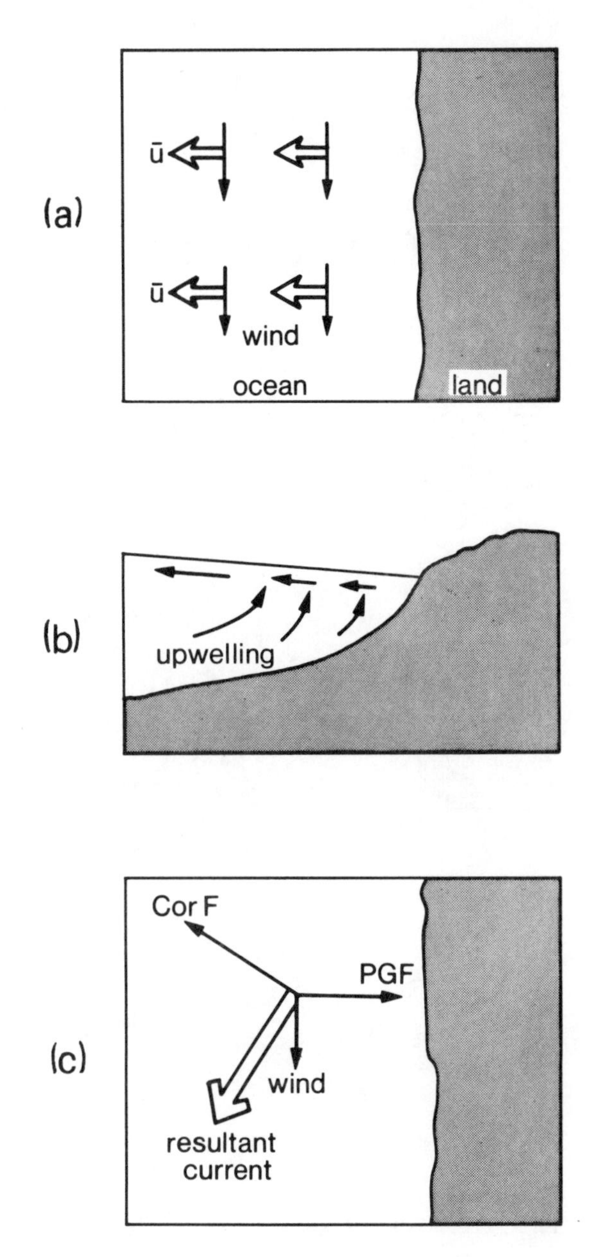

14. 6 Upwelling in eastern boundary currents in the northern hemisphere:

(a) Plan view showing Ekman transports resulting from the wind stress.

(b) Profile showing associated vertical water movements and sloping sea surface.

(c) Possible balance of forces and resultant current.

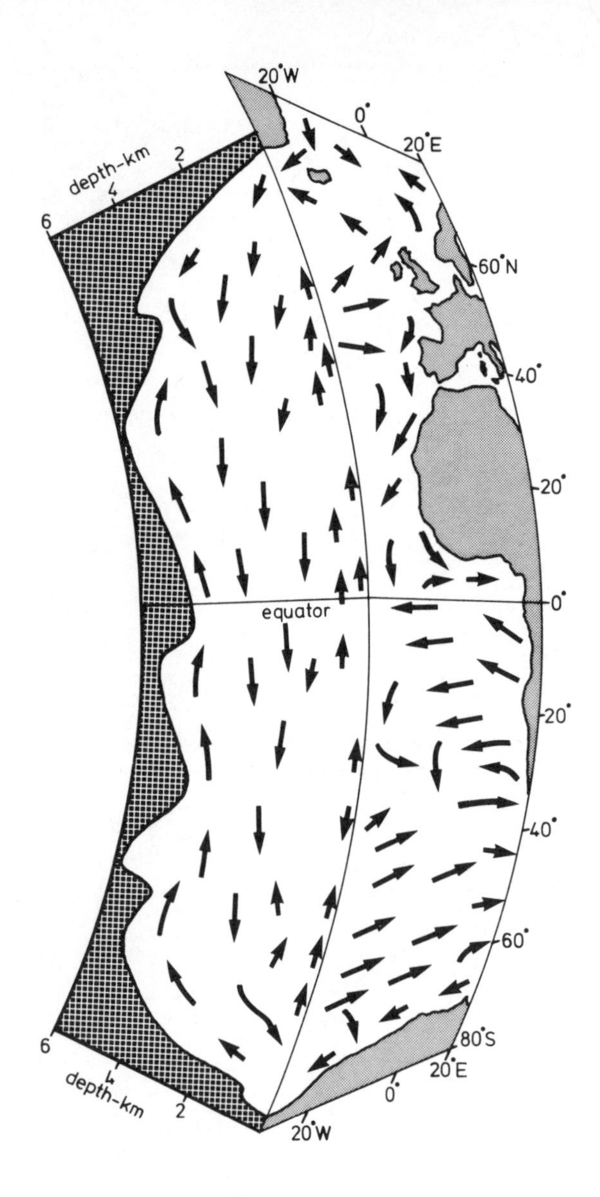

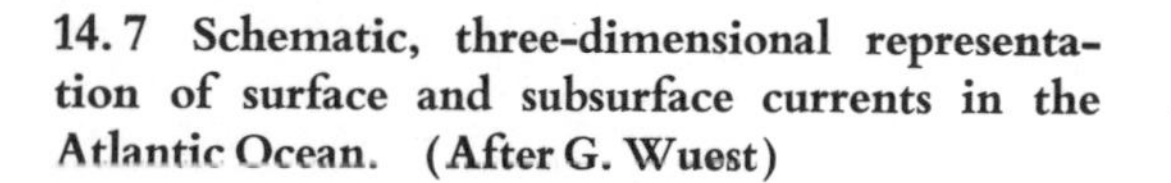

14. 7 Schematic, three-dimensional representation of surface and subsurface currents in the Atlantic Ocean. (After G. Wuest)

where the wind-induced transport is off-shore, and it requires upwelling from below. This brings water rich in nutrients to the photic zone, and hence areas of upwelling are characterised by considerable biological productivity. The main regions to which this applies are the cool eastern boundary currents of the subtropical gyre (Fig. 14. 6). The anomalously low temperatures of these currents are depressed further by the upwelling of cool water, and thus advection fogs are often encountered in such regions. The vertical water movements associated with the convergences and divergences so far discussed are chiefly limited to the water above and within the main thermocline. Theoretical considerations show that over *most* of the oceans water must well up slowly through the main thermocline. This conclusion may be reached either by considering the divergence in the flow above the main thermocline, or by considering the processes necessary to sustain the main thermocline in spite of the downward transport of heat by diffusion. Water sinks to offset this transport in only very limited areas of the world ocean determined by climatological factors. These areas may be located by water mass analysis, in particular by the study of dissolved oxygen concentration in the deep water masses.

Intermediate water masses which never sink to depths greater than about 2 000 m in the ocean originate, for example, at the Antarctic Convergence, and in the Mediterranean and Red Seas. The only source regions of deeper water masses lie around the Antarctic continent and in the North-East Atlantic near Greenland, though the water masses which sink in these regions are modified by mixing in their passage around the oceans. Figure 14. 7 gives an impression of the way in which these water masses are thought to move in the Atlantic Ocean, and the way in which they are related to the circulation of the upper layers. Considerations similar to those used to explain the intensification of the western boundary currents in the subtropical gyre show that deep water which is slowly upwelling will usually move towards the poles. It has therefore been postulated that there are deep, narrow western boundary currents in all of the oceans which are mainly responsible for transporting the deep water masses around the oceans, and that throughout the remainder of the oceans the deep water moves slowly polewards (Fig. 14. 8). But most of the evidence for these patterns is indirect, and there have been few observations to support them, so that Figures 14. 7 and 14. 8 should only be considered as portraying models of the deep circulation of the oceans which still require testing.

14.8 Model of deep-water circulation of the world ocean with source regions in the North Atlantic and in the Weddell Sea. (After H. Stommel)

Summary of Mathematical Expressions

General: the differential coefficient

$\Delta[Q]$	change of quantity [Q]	[Q]
$\frac{d[Q]}{dz}$	rate of change of quantity [Q] with depth	$[Q]\ m^{-1}$
$\frac{d[Q]}{dt}$	rate of change of quantity [Q] with time	$[Q]\ s^{-1}$

Chapter 2

e	mathematical constant	$= 2.718$
g	acceleration due to gravity	$\simeq 9.81\ m\ s^{-2}$
M	molecular weight	$kg\ mol^{-1}$
p	pressure	mb ($= 10^2\ N\ m^{-2}$)
R	universal gas constant	$= 8.3\ J\ mol^{-1}\ K^{-1}$
R'	R/M	$J\ kg^{-1}\ K^{-1}$
T	temperature	K, °C
v	volume	m^3
z	height or depth (positive upwards)	m
θ	potential temperature	K, °C
ρ	density	$kg\ m^{-3}$
ρ_w	density of water	$kg\ m^{-3}$
$\int \ldots$	integral of....	

Chapter 3

e	water vapour pressure	mb
e_w	saturated water vapour pressure	mb
U	relative humidity	% (per cent)

Chapter 4

S	salinity	‰ (parts per thousand)

Chapter 5

E	radiation emitted per unit surface area	$W\ m^{-2}$
I	incoming short-wave radiation	$W\ m^{-2}$
R	outgoing long-wave radiation	$W\ m^{-2}$

t	time		s
T	temperature		K, °C
σ	Stefan's constant	=	5.67×10^{-8} W m^{-2} K^{-4}

Chapter 6

ρ	density, specific gravity		g cm^{-3}, ---
σ	a function of specific gravity	=	$(\rho - 1)\ 10^3$
σ_t	value of σ at atmospheric pressure		
σ_θ	value of σ at atmospheric pressure and potential temperature		

Chapter 8

a	acceleration		m s^{-2}
f	Coriolis parameter	=	$(2\Omega \sin\phi)$ s^{-1}
F	force		N (= kg m s^{-2})
m	mass of body		kg
mv	momentum		kg m s^{-1}
r	radius of circle, distance from axis of rotation		m
δr	change in r		m
t	time		s
δt	time interval		s
u	speed		m s^{-1}, knots
$\vec{u}$	velocity		m s^{-1}, knots
$\vec{u}_i$	instantaneous velocity		m s^{-1}, knots
$\vec{u}_1, \vec{u}_2, \vec{u}_3, \vec{u}'$	vector components of velocity		m s^{-1}, knots
u_x, u_y	components of velocity in x and y directions		m s^{-1}, knots
x	distance towards east		m
y	distance towards north		m
δy	distance moved northwards		m
θ	direction of movement relative to north		rad, degrees
ϕ	latitude		rad, degrees
ω	angular speed of rotation		(rad) s^{-1}
Ω	angular speed of Earth's rotation	=	(7.29×10^{-5}) s^{-1}

Chapter 9

a	amplitude of wave	m
c	wave speed	m s^{-1}, knots
d	depth of water	m
T	period of wave	s
f	wave frequency ($= 1/T$)	s^{-1}
g	acceleration due to gravity	$\simeq$ 9.81 m s^{-2}
H	wave height ($2a$)	m
l	length of tank or basin	m
L	wavelength	m
t	time	s
x	distance in direction of wave movement	m
η	displacement due to wave	m

Chapter 10

m	mass	kg
p	pressure	mb
r	radius of curvature of isobars	m
u	speed	m s^{-1}
U_g	speed of geostrophic wind	m s^{-1}
U_{gr}	speed of gradient wind	m s^{-1}
U_{cyc}	speed of cyclostrophic wind	m s^{-1}
x, y, z	rectangular coordinate system	
$\delta x, \delta y, \delta z$	small distances in the x, y and z directions	m
α	angle between isobars and wind direction	rad, degrees
ϕ	latitude	rad, degrees
ρ	density	kg m^{-3}
Ω	angular speed of Earth's rotation	$= (7.29 \times 10^{-5})$ s^{-1}
$\frac{\partial p}{\partial x}$	rate of change of pressure with distance in the x direction, on the assumption that y and z remain constant (i.e. the partial differentiation coefficient)	mb m^{-1}

Chapter 11

f	Coriolis parameter	$= (2\Omega \sin\phi)$ s^{-1}
g	acceleration due to gravity	$\simeq$ 9.81 m s^{-2}
h_A, h_B	heights of water columns A and B	m
K	$(R'/g)(\Delta p/p)$	m K^{-1}
m	mass	kg
p	pressure	mb
R'	R/M (see Chap. 2)	J kg^{-1} K^{-1}

$\overline{T}$	mean temperature of layer	K, °C
u	geostrophic current speed	m s^{-1}
U_1	component of geostrophic current	m s^{-1}
U_{G1}, U_{G2}	components of geostrophic wind	m s^{-1}
U_T	thermal wind component	m s^{-1}
z	height or depth	m
θ	angle between horizontal and isobaric surfaces	rad, degrees
ρ_A, ρ_B	mean densities of water columns A and B	kg m^{-3}
ϕ	latitude	rad, degrees
Ω	angular speed of Earth's rotation	$= (7.29 \times 10^{-5})$ s^{-1}

Chapter 12

f	planetary vorticity (= Coriolis parameter)	s^{-1}
ζ	relative vorticity	s^{-1}

Chapter 14

f	Coriolis parameter	$= (2\Omega \sin\phi)$ s^{-1}
D	depth frictional influence of wind	m
$\overline{u}$	depth mean current in Ekman layer	m s^{-1}
ρ	density	kg m^{-3}
τ	wind stress on water surface	N m^{-2}

List of Further Reading

The Atmosphere

R. G. BARRY and R. J. CHORLEY Atmosphere, Weather and Climate. (*Methuen, 2nd Ed. 1971, 379 pp.*)
Intended specifically for geography students, and uses the approaches of synoptic and dynamic climatology to provide an understanding of climate and weather through a knowledge of the workings of the atmosphere. Includes a fairly comprehensive bibliography.

H. R. BYERS General Meteorology. (*McGraw-Hill, 4th Ed. 1974, 416 pp.*)
A classical textbook giving a clear exposition of the fundamentals of meteorology for the student who is reasonably conversant with mathematics.

W. L. DONN Meteorology. (*McGraw-Hill, 4th Ed. 1975, 518 pp.*)
A general introductory text on meteorology, with particular emphasis on the marine applications of the subject.

H. H. LAMB Climate: Present, Past and Future; Vol 1: Fundamentals and Climate Now. (*Methuen 1972, 613 pp.*)
Presents and explains the present-day world distribution of climates to provide a basis for comparison with the features that have characterised other climatic epochs. Pays particular attention to the effect of the ocean on climate.

J. G. LOCKWOOD World Climatology – an Environmental Approach. (*Arnold 1974, 330 pp.*)
Describes the climates of each of the world's major regional zones, emphasising the aspects which are ecologically important.

D. H. MCINTOSH and A. S. THOM Essentials of Meteorology. (*Wykeham 1969, 238 pp.*)
A good introduction to the subject for those with mathematics and physics up to university entrance standard. Provides a particularly helpful table of symbols, units and numerical values, and problems for the student to attempt at the end of most chapters.

S. PETTERSSEN Introduction to Meteorology. (*McGraw-Hill, 3rd Ed. 1969, 333 pp.*)
A general introductory text, making very sparing use of mathematics, and relying rather on physical explanation in the text well illustrated by diagrams.

E. T. STRINGER Foundations of Climatology. (*Freeman 1972, 586 pp.*)
A fairly comprehensive introduction to physical, dynamical, synoptic and geographical climatology. Written from the viewpoint of classical physics, but suitable for the non-mathematical reader.

The Ocean

W. S. von ARX An Introduction to Physical Oceanography. (*Addison-Wesley 1962, 422 pp.*)
Principally but not solely concerned with water movements in the ocean and their investigation. Provides an introduction to some of the basics of fluid mechanics, but generally uses physical rather than mathematical explanations of oceanic phenomena.

P. GROEN The Waters of the Sea. (*Van Nostrand 1967, 328 pp.*)
Mainly descriptive, but gives some account of the physical processes at work in the marine environment. Pays considerable attention to ice in the seas.

M. G. GROSS Oceanography, a View of the Earth. (*Prentice-Hall 1972, 581 pp.*)
Takes a broad view of the world ocean as the Earth's most distinctive feature, and examines the processes controlling its major features and the life in it.

H. J. MCCLELLAN Elements of Physical Oceanography. (*Pergamon 1965, 151 pp.*)
Comprises a brief account of descriptive physical oceanography and a mainly mathematical approach to oceanic movements, followed by rather more detailed accounts of a few selected topics.

G. NEUMANN and W. J. PIERSON Principles of Physical Oceanography. (*Prentice-Hall 1966, 545 pp.*)
For the advanced student of physical oceanography; a fairly comprehensive text with an extended list of references to original literature.

G. L. PICKARD Descriptive Physical Oceanography. (*Pergamon, 2nd Ed. 1975, 214 pp.*)
An excellent introduction to the synoptic aspects of physical oceanography, in particular the circulation and water masses of the oceans, but the reader will have to turn to other texts for consideration of the dynamics of the phenomena described.

P. WEYL Oceanography, an Introduction to the Marine Environment. (*Wiley 1970, 535 pp.*)
An ambitious text dealing with physical, geological, chemical and biological aspects of the marine environment and trying to set them into the wider context of man's total environment.

The Atmosphere and Human Activity

W. J. MAUNDER The Value of the Weather. (*Methuen 1970, 388 pp.*)
Shows how man is affected by weather and how he reacts to it, and then attempts an economic analysis of weather and of the benefits and costs of weather knowledge.

K. SMITH Principles of Applied Climatology. (*McGraw-Hill 1975, 233 pp.*)
Reviews a wide range of literature concerned with the relationships between the atmospheric environment and human activity.

J. A. TAYLOR (ED.) Climatic Resources and Economic Activity. (*David & Charles 1974, 262 pp.*)
A rather diverse collection of papers presented at a symposium in 1972, providing a series of case studies of the impact of weather on man's activities.

The Ocean and Human Activity

D. W. HOOD (ED.) Impingement of Man on the Ocean. (*Wiley 1971, 738 pp.*)
A collection of chapters written by leading experts to show how the ocean works, and the effects of contaminants and exploitation on the ocean's processes.

M. M. SIBTHORP (ED.) The North Sea, Challenge and Opportunity. (*Europa 1975, 324 pp.*)
The report of a study group on man's uses of this particular shelf sea, and the ways in which the present legal regime affects these uses, particularly where there is a conflict between them.

B. J. SKINNER and K. K. TUREKIAN Man and the Ocean. (*Prentice-Hall 1973, 149 pp.*)
A brief account of the resources of the ocean, and the factors which control their distribution and exploitation, and an even briefer account of the pollution of the oceans.

Interaction of Atmosphere and Ocean

E. B. KRAUS Atmosphere-Ocean Interaction. (*Oxford U. P. 1972, 275 pp.*)
An advanced text for the student who is already familiar with either meteorology or physical oceanography, dealing with the nature of the two fluids near their interface and the exchanges of energy and momentum between them.

INDEX

Bold page numbers indicate the position of a chapter which deals largely with the corresponding subject. *Italic* page numbers indicate the location of a titled sub-section of a chapter which deals *in detail* with the corresponding subject.

Acceleration, in laws of motion, 65, 68
Adiabatic changes, *17*, 25, 51, 52–3
Advection of heat, 45, 50
Advection fog, 24, 110, 126
Africa, West, monsoon area, 107
Air
adiabatic changes, 17, 25, 52–3; density, 18, 52–3; effect of behaviour on cloud formation, 26, 117; effect on, of laminar, turbulent flow, *66–7*; moist, relation to tropical cyclones, 114; parameters of water vapour in, 22–3; rising, in atmosphere, 25–6; saturation, condensation, 23–4; *temperatures*: associated with anticyclones, 116, diurnal, seasonal, 45–50, determining factors, 50, potential, *in situ*, 17, relation to formation of sea ice, 35–7; *see also* Air Masses, Atmosphere
Air masses
classification, properties, sources, 58, *59–60*; concept, *59*; frontal uplift, 26; modification, *60*; over Britain, 58, *60–1*, 67; *see also* Atmosphere
Alaska Current, 125
Albedos, of various surfaces, clouds, 41–2, 42–3, 47
Altitude, of land surface, effect on air temperature, 50; *see also* Orographic barriers
Altocumulus clouds, 26–8
Altostratus clouds, 26–8, 110
Amphidromic tidal systems, 82
Amplitude, in wave motion, 73–4
Anabatic winds, *86–7*
Anemometer, 64
Antarctic Circumpolar Current, 124
Antarctic Convergence, 124, 126
Antarctic Intermediate Water, 124
Antarctic region
air mass source, 59–60; albedo, 47; ice, icebergs, 30, 35–6, 36–7, 125; ocean basin, 12; water masses, currents originating in, 62–3, 126, 127
Anticyclones, 89–90, 105–6, **109–16**, 121
Anvil head, 27
Arabian Sea, tropical cyclones in, 115
Arctic regions
air mass sources, 59–60; effect on British weather, 61; ice, icebergs originating in, 30, 35–6, 37, 125; ocean basin, 12; water masses, currents originating in, 62–3, 125
Arctic sea smoke (steam fog), 24
Atlantic Ocean
air temperatures, depressions over, 50, 106; basin, 10, 12; currents, water masses, 62–3, 121–5, 126, 127; ice, icebergs, 37, 125; nutrient salts, 35; pressure belts, wind systems, 99, 115; thermal layers, 56
Atmosphere
absorption, transmission of radiation, *40–2*, 43, 56–7; air rising in, 25–6; circulation systems, **99–107**; classification, types of motion in, 66–71, 82; composition, structure, 11, 15–20, 34, 43, 56–7; gases from, dissolved in sea water, 33–4, 35; heat transfer to, from oceans, *44–5*; laminar, turbulent flow in, *67–8*; lower, heat source for, 53–4; origins, **11–13**; residence time of water in, 21–2; response to various forces, *67–8*; significance of Coriolis Force, *69–71*; stability in, *52–3*, 54; *temperatures*: inversion, 55, 67, potential, *in situ*, 17, relation to isobaric layer thickness, 96–7, upper, *56–7*; *see also* following entry.
Atmospheric pressure
determining factors, 15–16, 17, 22–3, 52–3; diurnal, tidal variations, 82; idealised distribution in northern hemisphere, 105; levels associated with various weather types, 106–7, 110, 111, 113–16; surface, and wind patterns, 99–100; *variations*: effect on ocean currents, 120–1, in mid-latitudes, 105–6; zonal belts and wind systems, 99–100, *106–7*
Australia, north, monsoon area, 107

Baltic Sea, water masses, 62
Barotrophic conditions, 93–4
Bay of Bengal, cyclones, 115
Bay of Fundy, spring tides, 81
Bays, waves, currents in, *78–9*, 81–2
Beachy Head, 72
Beaufort Wind Scale, 75, 76, 113
Bergen Model, 109, 110
Bergeron-Findeison Theory, 29
Bora, katabatic wind, 87
Breezes, land, sea, 67, *85–6*, 99, 116
Britain, weather patterns, 58, *60–1*, 67, 105
Buys-Ballot's Law, 88

Calcium, levels in sea water, 31
Capillary waves, 20
Carbon dioxide, in atmosphere, 11, 17, 34, 43
Central Water Masses, 62
Centrifugal force (CenF), in Earth's rotation, *68*, 69, 79–80, 89–91
Centripetal acceleration, force, 68
China Sea, monsoon area, 107
Chinook winds, 53
Chlorinity titration of water, 31–2
Circulation systems
  atmospheric, **99–107**; oceanic, **119–27**; *see also* Atmosphere, Currents, Oceans, Winds
Cirrocumulus clouds, 28
Cirrostratus clouds, 26, 114
Cirrus clouds, 26–8, 104, 109–10, 114
Clouds
  absorption, reflection of radiation, 40–2, 43; in hydrological cycle, 21–2, *25–6*, *26–8*; residence time of water in, 21–2; types described, *26–8*; water droplets, 28, 29; weather associated with, 25–8, 41–2, 53, 55, 56, 61, 86, 87, 91, 104, 109–11, 114, 116
Coastlines, erosion of, 78, 81
Compressibility, of fluids, *15–16*, 17, 52
Compression, in wave motion, 73
Condensation
  in air, 52–3; in hydrological cycle, *23–4*; in transfer of heat between ocean and atmosphere, 44–5; nuclei, 23
Conduction *see* Heat
Continental regions
  air mass sources, 59–60; *effects*: on atmospheric circulation, *106–7*, on British weather, 61
Continental shelves, 11–13, 82
Convection
  in atmosphere, effects, 26, 28, 44, 55, 87; of heat, by Earth-surface factors, 42–3; turbulent, in heat transfer, 44; *see also* Heat
Coriolis Effect, 65–6
Coriolis Force (CorF), 66, 68, *69–71*, 82, 86, 87–8, 89–91, 93–4, 95, 100, 107, 119–20, 123, 124
Cumuliform clouds, 28
Cumulonimbus clouds, 28, 56, 61, 91, 111, 114
Cumulus clouds, 27, 28, 53, 55, 61, 86, 110, 111
Current meter, 64
Currents, ocean
  deep water: and vertical movements, 125–7, associated with surface circulation, 121, source regions, *125–7*; effect of gyres, 121–2, 125, 126; expression of, 65; geostrophic, **93–8**, 123–4; influence on icebergs, 37, 125; interacting causes, 57, 119, 120–1, 123–4; relation to winds, 57, 67, *76–7*, 95, *119–21*, 124; seasonally reversing, 67; surface, *121–5*; tidal, 82; transmission of heat by, 42, 50; turbidity, 13, 31; upwelling in northern hemisphere, 125; velocity, 67
Cyclogenesis, 113
Cyclones
  causes, effects, 26, 89–90, 105–6, **109–16**; in scales of motion, 67; mid-latitude, 105–6, *109–13*; tropical, *113–16*, 120–1; vorticity, 104
Cyclostrophic winds, *91*

Day, civil, sidereal, distinguished, 88
Density
  distribution, an internal force, 68; of air, determining factors, 52–3; of ice, sea water, fresh water, 18–19; of fluids, sea water, relation to other factors, 15, 18, 35, 51–2, 93, 94–6
Depressions
  associated weather, 109–11, 117; lee, thermal lows, polar air, 113; mid-latitude, 67, 105–6, *109–13*, 120–1; on polar front, 113; secondary, 113; tropical, *113–16*; *see also* Cyclones
Depth, ocean
  relation to: behaviour of currents, 94–6, 119–21, *125–7*, temperature, 46, 47, 57, 61–2, water constituents, 32–3, 33–4, 35, 61–2
Dew, 23, 24
Dewpoint, 23
Direction, in laws of motion, 65; *see also* Wind
Dishpan experiments, *100*, 101, 104
Displacement, in wave motion, 73, 74
Distance, in wave motion, 73, 74
Dry Adiabatic Lapse Rate (DALR), 52–3, 56
Dye tank, 118

Earth
  evolution, **11–13**; incoming, outgoing radiation, 24, 39, *42–3*, *43–4*, 50, 53–4; planetary temperature, 43; *significance of*: rotation, 26, 65–6, 68, *69–71*, *79–80*, 100, 104, surface factors, 24, *42–3*, 50, 53–4, 88–9

Earthquakes, 13, 73, 78
East Greenland Current, 125
Easterly wind systems, 114; *see also* Trade winds
Ekman spiral, *119–21*, 124, 125
Energy
distribution, *39–40*, *40–2*, *45–50*; in seas, dissipation on headlands, 78; in waves, relation to height, 76, 77; storage in water, 20; *see also* Heat, Radiation
'Eye' of storm, 108

Falkland Current, 125
Fetch of sea, defined, 76
Flow, laminar, turbulent
effect on viscosity, 18; in atmosphere, oceans, 57, *66–7*; within clouds, 29
Fluids
adiabatic changes, *17*, 51; composition, properties, **15–20**; compressibility, *15–16*, 17; defining properties, 15; potential temperature, 17; relation of density, viscosity, 18; response to flow, *66–7*; stability, *51*; *see also* Sea water, Water
Fog, 24, 25–6, 40–1, 45, 55, 61, 86, 110, 116, 126
Föhn winds, *53*, 55
Food chains, oceanic, 32–3
Force
centrifugal, in Earth's rotation, *68*, 69, 79–80, 89–91; centripetal, 68; external, internal, secondary, distinguished, 68; *see also* Coriolis Force
Franz Josef Land, icebergs from, 36
Frazil ice, 35
Freezing points, 18, 19, 35–7
Frequency, in wave motion, 74, 77
Friction, effects of
on acceleration of sea water, 95; on motion, 68; on ocean currents, 119–21; on wind speed, direction, 88–9, 90, 91
Frontal (mixing) fog, 24
Frontal uplift
of air masses, 26; types of cloud formed by, 28
Fronts, 59, 109–10, 110–11, 112, 117
Frost, 23, 24, 61, 116; *see also* Ice
Frost-point temperature, 23

Gases
dissolved, in sea water, *33–4*, 35; inter-relation of density, compressibility, 15
Geostrophic currents
and thermal winds, **93–8**; causes, effects, 123–4; gradient equations, *93–4*; Helland-Hansen's Equation, *94–5*; problems in calculation of, 94–6; relation to density, depth, 94–6; speeds of geostrophic winds and, compared, 93
Geostrophic winds
causes, effects, *87–8*; speeds of geostrophic currents and, compared, 93
Glycerine, density, viscosity, 18
Gradient equations, *93–4*
Gradient winds, *89–90*
Graham Land, 28
Grand Banks, Newfoundland, 13, 24, 25, 45 125
Gravitational pull
balance in Earth–Moon system, *79–80*; in gradient equations, 93–4; of Earth, effect on vertical motion, 65; of Sun, Moon, effect on tides, 68; waves dependent on, 73
Greenhouse effect, 43
Greenland, icebergs from, 36, 125
Groundwater, residence time of water in, 21, 22
'Growlers', 37
Gulf Stream, 24, 50, 67, 121, 124, 125
'Guyots', 13
Gyres, subpolar, subtropical, 121–2, 125, 126

Hadley cell, *100*, 101, 102, 107
Hail, 29; *see also* Precipitation
Halo, round Sun, Moon, 26
Harbours, standing waves in, 79
Haze, 24, 40–1; *see also* Fog
Headlands, erosion of, 78
Heat
advection, 45, 50; balance, and back radiation, *43–4*, 61; *distribution*: by Earth's surface, 42–3, 44, by solar radiation, *39–40*, downward, in thermoclines, 57, significance of temperature variation in, *45–50*; sources for lower atmosphere, 53–4; specific, latent, of water, 20, 42–3, 46; transfer between atmosphere, oceans, *44–5*; transport polewards, 99–100, 101
H.M.S. Hecla, 14
Height
of atmosphere, inter-relation with other factors, 15–16, 16–17, 52–3, 53–5, 56–7, 100–4; *of waves*: in Beaufort Scale, 75, inter-relation with other characteristics, 73–4, 76, 77, 78; *tsunamis*, 78
Helland-Hansen's Equation, *94–5*
Hill fog, 24, 25–6
Hills *see* Mountainous regions, Orographic barriers
Hoar frost, 23
Hooked cirrus clouds, 26
Horizontal pressure gradients *see* Pressure gradients
Humidity, of air, 23, 53, 114; *see also* Water vapour
Hurricanes, 108, *113–16*
Hydrological cycle, **21–9**
Hygroscopic condensation nuclei, 23, 24
Hypolimnion, in fresh water, 57

Ice
crystals, in clouds, 26, 28, 29; formation, melting, types of, 20, 35–7; in oceans, *35–7*; inter-relation of density, temperature, 18–19; residence time of water in, 21; sublimation, 22, 23; *see also* Icebergs
Icebergs
boundaries, distribution, 36, 37, 125; formation, composition, *35–7*; monitoring of, 37; northern, southern, distinguished, 30, 36–7
Ice-blink, 35–6
Ice islands, 36
Iceland, low pressure area, 106
Indian Ocean
basin, 12; currents, 67, 124, 125; monsoon area, 107, 125
Indian sub-continent, monsoon area, 67, 106–7
Instability, in fluids, 51
Intermediate Water Masses, 62
Internal waves, in oceans, 53
International Cloud Atlas, 28
Inversion layers, in atmosphere, 55, 67
Ionosphere, 40, 57
Irminger Current, 125
Isobars, 85, 96–7
Isopycnals, 62
Isopycnic surfaces, 94
Isotherms, 50

Jet Streams, 27, *100–5*, 107

Katabatic winds, *86–7*
Kuroshio area, varying water temperatures, 47
Kuroshio Current, 121

Labrador Current, 24, 125
Lakes
inverse thermoclines in, 57; residence time of water in, 21; standing waves, *78–9*
Land
breezes, 67, *85–6*, 99, 116; evaporation of water from, 22–3; effect on atmospheric circulation of seasonal factors, 106–7; rates of cooling of sea and, compared, 24; relation of surface to air temperature, 50; surface albedos, 42–3; *see also* Earth
Latitude
relation to: constituents of sea water, 32–3, 33–4, 35, 36, development of anticyclones, 116, distribution of wind belts, 99, formation of tropical cyclones, 114, 115, function of CorF, 86, occurrence of Hadley Cell, 100, 102, reception of solar radiation, 40, 41, 42, 44–5, surface temperature, *47–50*, thermoclines, 57
Lee waves
in air, 53; in cloud formation, 25–6; in scales of motion, 67
Lenticular clouds, 25–6
Longitudinal waves, 73

Maritime regions
air mass sources, 59–60; effect on British weather, 61
Mass, in laws of motion, 65
Masses *see* Air masses, Water masses
Mediterranean Sea, water masses, 62, 126
Mesopause, 57
Mesosphere, 57
Mid-Atlantic Ridge, 10, 12
Mid-latitudes *see* Depressions, Westerly winds
Mist, 24; *see also* Fog
Mistral, 62, 87
Mixing (frontal) fog, 24
Monsoons, 67, *106–7*, 125
Moon
gravitational pull, 68, 79–80; halo, 26
Motion
and forces, **65–71**; classification, types, *66–71*; effect on viscosity, 18; effects of turbulent flow, 29, 67; horizontal, vertical, determining factors, 65–6, 68, *69–71*; in vortices, and polewards transport of heat, 99–100, 101; Laws of, *65–6*; scales of, *67*; seasonal, 67; significance of friction, 68; *see also* Particle motion, Waves
Mountainous regions
effect on atmospheric behaviour, 25–6, 53, 54–5, 106–7; weather associated with, 24, 25–6, 28, 61, *86–7*, 110

Newfoundland, Grand Banks, 13, 24, 25, 45, 125
Nimbostratus clouds, 26, 110
Nitrates, in sea water, 32–3, 34
Nitrogen, in atmosphere, 11, 17
North Atlantic Drift, 50
Northern hemisphere
effect of Coriolis Force, *69–71*; factors determining nature of winds in, 87, 88, 89, 97, 100, 114, 115, 116; idealised pressure distribution, 105; positive vorticity, 104; summer, winter atmospheric circulation compared, 106; surface currents, 121–5; upwelling in eastern boundary currents, 125
North Sea
storm surges, 121; tidal system, 82
Novaya Zemla, icebergs from, 36
Nutrient salts, *32–3*, 62

Occlusions, 111, 113
Ocean basins, 10–13, 81
Oceanic crust, 11–12
Ocean ridges, 10
Oceanic stratosphere, concept of, 57

Oceanic trenches, 12–13
Oceanic troposphere, concept of, 57
Oceans
Beaufort Scale of surface state, 75, 76; classification, types of motion in, *67–71*; currents, water masses, 13, 31, 37, 42, 50, 57, *61–2*, *62–3*, 65, 67, 82, **93–8**, **119–27**; density, compressibility, 15–16; fluid components, properties, **15–20**; internal waves, 53; interrelation of depth, temperature, and other factors, 17, 32–3, 33–4, 35, 46, *47–50*, 57, 61–2, 94–6, 119–21, *125–7*; origins, **11–13**; photic zones, 32–3, 33–4; postulated layers, 57; pressure gradients, 85, *93*, 94, 95–6; rates of surface cooling of land and, compared, 24; relation of distribution of land and, to air temperature, 50; residence time of water in, 21; response to force, *67–8*; rôle in heat transfer, 42–3, *44–5*, 106; salts, gases, ice in, **31–7**; stability, *52*; sloping surface, 93, 94, 121–5; surface albedos, 42–3; thermoclines, *57*; *see also* Icebergs, Sea ice, Sea water, Tides, Waves
Orographic barriers, 25–6, 28, 53, 55, 61, 110
Oxygen
in atmosphere, 11, 17, 40; in evolutionary history, 11; in sea water, 33–4, 35, 62
Oyashio Current, 125
Ozone, 40, 43, 56–7

Pacific Ocean
basin, 12; constituents, 33, 34; currents, 121–5; pressure belts, wind systems, 99, 115; surface temperatures, 57; *tsunamis*, 78
Pack-ice, 35–6
Pancake ice, 35
Particle motion, *77–8*
Period, in wave motion, 73, 76, 77, 78–9, 80–2
Peru Current, 122
Phosphates, in sea water, 32–3, 34, 62
Photic zones, 32–3, 33–4
Planetary temperature of Earth, 43
Plate tectonics, 12
Polar ice, 35
Polar regions
air mass sources, 59–60, 113; influence on British weather, 61; transport of heat towards, 99–100, 101; *see also* Antarctic, Arctic
Precipitation
and Föhn effect, 53, 55; effect on salinity of ice, 35; formation of water droplets, 29; from different types of cloud, 26–8; in hydrological cycle, *29*; weather types associated with, 55, 61, 110, 111, 112, 114, 116
Pressure *see* Atmospheric pressure, Pressure gradients, Water pressure
Pressure gradient force (PGF), 86, 87–8, 89–91, 95–6, 107
Pressure gradients, horizontal
and associated winds, **85–91**; effect on land, sea breezes, *85–6*; equations, *93–4*; estimation of force, *86*; in oceans, atmosphere, 85, *93*; interaction with CorF, 87–8, 89–91
Prominence, solar, 38

Radiation, solar, 38; terrestrial
and air temperature on mountains, 54–5; and temperature inversions, 55; balance, between Earth, atmosphere, *43–4*, *45–6*, 61; character, methods of measurement, *39–40*; *distribution*: effect of Earth's surface, *42–3*, significance of temperature variations, *45–50*; factors affecting absorption, 40, 42–4, 54, 56–7; in distribution of energy, heat, *39–40*; relation of reception, to latitude, season, 40, 41, 42; transmission through atmosphere, *40–2*
Radiation fog, 24, 86, 116
Rain *see* Precipitation
Rarefaction, in wave motion, 73
Red Sea, water masses from, 126
Refraction, in wave motion, 77, 78, 81
Residence time, of water in stages of hydrological cycle, 21–2
Resonance, of tide-producing forces, 80–2
Ridges, in Rossby waves, 104, 105
Ripples, nature of, 73
Rivers, residence time of water in, 21
Rock, rates of temperature increase in water and, compared, 42–3
Rocky Mountains, domination of weather by, 106; *see also* Mountainous regions
Ross Sea, ice, water masses from, 36, 125
Rossby waves, *100–5*, 116

Salinity
causes, effects, 19, *31–2*, 33, 35, 36, 52, 61; in identifying water masses, *61–2*, 63
Salts, nutrient, *32–3*, 62
Satellite photographs, 108, 117
Saturated Adiabatic Lapse Rate (SALR), 52–3; *see also* Adiabatic changes
Saturation vapour pressure, 23–4, 29, 45
Scattering of radiation, 40–1, 42–3
Sea breezes, causes effects, 67, *85–6*, 99
Sea-floor spreading, 10, 12, 13
Sea fog, 24
Sea-ice, 35–7; *see also* Icebergs
Sea level, determining factors, 12
Seamounts, 13
Sea water
adiabatic changes, 17; Beaufort Scale on surface state, 75, 76; chlorinity titration of, 31–2; compressibility, 52; constituents, **31–7**; dissolved gases, *33–4*; in sea-ice, icebergs, *35–7*; nutrient salts, *32–3*, 62; salinity, *31–2*, 33, 35, 36, 52, *61–2*, 63; *significance of*: density, 15, 18, 35, 51–2, 93, 94–6, depth, 32–3, 33–4, 35, 46, *47–50*, 57,

61–2, 94–6, 119–21, *125–7*, latitude, 32–3, 33–4, 35, 36, *47–50*, temperature, 46, *47–50*, 61; vertical movement and deep circulation, *125–7*; *see also* Fluids, Oceans
Seiches, 73, *78–9*, 81–2
Seismic sea waves, 78
Shallow water effects, *77–8*
Silicates, in ocean ecology, 32–3, 34, 62
Sinusoidal waves, 73, 74
Sky, colour of, 40–1
Sleet, 29, 61; *see also* Precipitation
Smog, 24, 116; *see also* Fog
Snow, 29, 61; *see also* Precipitation
Solar constant, 40
Sound waves, 73
Southern hemisphere
cyclonic, anticyclonic situations, 89–90; depressions in, 111; effect of Coriolis Force, 69–71; factors determining nature of winds in, 88, 97, 100, 114, 115, 116; geostrophic currents, 95; negative vorticity, 104; surface currents, 122, 124; *see also* Cyclones, tropical
Southern Ocean *see* Antarctic region
Speed
in laws of motion, 65; of ocean currents, 119–21; of waves, 73, 76, 77, 78, 81; of winds, 53, 55, 75, 76, 85–91, 93, 96–7, 103, 104–5, 113
Spindrift, 71
Spitzbergen, icebergs from, 30, 36
Stability
in fluids, oceans, *51*, *52*; in the atmosphere, *52–3*, 54; vertical, and temperature distribution, 16, **51–7**
Standing waves, 73, *78–9*, 81–2
Steam fog, 24
Steepness, in wave motion, 73, 77
Stefan's Law, 39, 43, 46
Storm surges, *120–1*
Storms
cyclostrophic, 91; 'eyes', 108; in scales of motion, 67; *see also* Beaufort Scale, Cyclones
Straits of Dover, currents, 94
Stratiform clouds, 109
Stratocumulus clouds, 26–8, 41–2, 61
Stratopause, 56–7
Stratosphere, 56
Stratosphere, oceanic, concept of, 57
Stratus clouds, 26–8, 53, 55, 61, 110
Sublimation, 22, 23
Sub-polar gyre, 125
Subtropical gyre, 121–2, 126
Sun
character of energy source, *39–40*; gravitational pull, 68, *79–80*; halo, 26; significance of angle of rays, 42, 50; sunspots, 38; *see also* Radiation
Surging, of standing waves, 79
Survey ship, 14
Sweeping, in raindrop formation, 29
Swell, distances travelled by, 77
Temperature
distribution, vertical stability and, 16, **51–7**; effect of variations on radiation balance, *45–6*; inversions, 55, 57, 116; of air, inter-relation with other factors, 23, 24, 25–6, *47–50*, 52–3, 53–5, 56, 61 96–7, 99; of air and water, in transfer of heat from ocean to atmosphere, 44–5; of Earth, effect of heat balance, back radiation, *43–4*; of fluids, sea water, inter-relation with other factors, 17, 18–19, 32, 35–7, 52, 57, *61–2*, 63; relative rates of increase in water, rock, 42–3; *structure*: of oceans, thermoclines and, *57*, of upper atmosphere, *56–7*; weather types associated with various levels, 110, 111, 114, 116
Temperature/Salinity gauge, 92
Tephigrams, 53, 54
Thermal equator, 47, 49
Thermal winds, *96–7*, 100–4
Thermoclines
in fresh water, 57; *relation to*: areas of upwelling, 126, laminar flow, 67, surface currents, 121–5, temperature structure of oceans, *57*
Thermosphere, 57
Thunderstorms, 55, 61, 111, 114
Tides
amphidromic systems, 82; diurnal, semi-diurnal periods, 80, 81; equilibrium, *79–80*; forces producing, 68, 80–1; frictional dissipation of energy from, 39; in scales of motion, 67; neap, spring, 80–1; observed, *80–2*; *see also* Oceans
Topographic lapse rate, 54–5
Tornados, 91
Trade winds, 67, 100, 107, 114, 122–3
Transverse waves, 73
Trochoidal waves, 74
Tropical regions
air mass sources, 59–60; cyclones in, *113–16*; effect on British weather, 61; formation of raindrops in, 29
Tropopause, 53–5, 56
Troposphere, *53–5*, 56, 100, 101, 104, 106
Troposphere, oceanic, concept of, 57
Tropospheric discontinuity, oceanic, concept of, 57
Troughs
in formation of cyclones, 114; in Rossby waves, 104, 105
T.S.D. gauge, 92
T-S diagrams, *61–2*, 63; *see also* Salinity, Temperature
Tsunamis, *78*
Turbidity currents, 13, 31
Typhoons, *113–16*

Upslope fog, 24, 25–6
Upwelling, areas of, *125–6*

Vapour pressure, of air, 45
Velocity
estimating, 67; in laws of motion, 65; in movement of air, water, 66; *see also* Coriolis Force
Viscosity, of fluids, 18–19
Visibility, factors affecting, 61, 110, 111, 116; *see also* Fog
Volcanic activity, areas of, 13
Vorticity, vortices, 99–100, 101, 104, 121–2

Water
condensation, 23–4; *deep*: short waves in, 74, particle motion in, *77–8*; density, viscosity, 18–19; effect of dissolved salt, 19; evaporation, transpiration, sublimation, *22–3*; freezing, boiling points, 18; fresh, behaviour of, 21, 57, *78–9*; in evolutionary history, 11; in the atmosphere, oceans, *18–20*; juvenile 22; molecular structure, 19, 20; rates of temperature increase in rock and, 42–3; residence times, during hydrological cycle, 21–2; *shallow*: long waves in, 74, particle motion in, *77–8*; specific, latent heat, 20, 42–3, 46; surface albedos, 42–3; surface tension, 20; *see also* Sea water, Water masses, Water pressure
Water masses
and T-S diagrams, *61–2*, 63; examples, *62–3*; properties, characteristics, *62*, 63, 118; source regions, 62–3, *125–7*; *see also* Currents, Oceans
Water pressure
changes, following vertical displacement, 51; in calculating salinity, 32
Water spouts, 84, 91
Water vapour
dewpoint, frostpoint, 23; formation of, 20; *in air, atmosphere*: absorption of radiation by, 40, 42, 43, 54, concentration 18, effect on density, 52, 53, parameters of, 22–3, residence time of water in, 21–2, saturation vapour pressure, 23; in evolutionary history, 11
Wave energy spectrum, 76, 77
Wavelengths, 73–4, 77, 78, 104–5
Waves
breaking of, 77; caused by wind, 67, *76–7*; coastline erosion by, 78; deep water, 74, *77–8*; during tropical cyclones, 115–16; essential parameters, 73–4; field formation, 76; *in air*: cause of cyclones, 114, lee, 25–6, 53, 67; *in oceans*: Beaufort Scale, 75, 76, interacting characteristics, 73–4, 75, 76, 77, 78, 81, internal, 53, particle motion and shallow water effects, 72, 74, *77–8*, progressive, 73–4, 81, response to tide-producing forces, 81, significance of surface tension of water, 20, standing, 73, *78–9*, 81–2, transmission of heat by, 42, 99–100, 101, tidal, 78, *tsunamis*, 78; see also Oceans, Rossby Waves, Sea water, Tides
Weather
British, 58, 60–1, 67, 105; forecasting, 59, 111–13; *see also individual types, aspects of weather*
Weddell Sea, icebergs, water masses from, 36, 127
Westerly wind systems
characteristics of upper, *100–5*; mid-latitude, *105–6*
West Greenland Current, 125
Winds
advection of heat by, 50; Beaufort Scale, 75, 76, 113; belts, determining factors, 99; *effects on*: cloud type, 26, 28, oceans, currents, 57, 67, *76–7*, 95, *119–21*, 124, transfer of heat from ocean to atmosphere, 45; factors affecting speed, direction, 53, 55, 75, 76, **85–91**, 93, 96–7, 103, 104–5, 113; geostrophic, *87–8*, 93, 96–7, 100–4; gradient, *89–90*, 100–4; gustiness, 67; influence of continental areas on, *106–7*; land and sea breezes, 67, *85–6*, 99, 116; patterns, and atmospheric pressure, *99–100*; pressure gradients and, **85–91**; relation to condensation, 24; surface, causes, effects, *88–9*; thermal, *96–7*, 100–4; weather associated with various types, 33, 55, 67, *86–7*, *105–6*, 109, 110, 111, *113–16*
Winter ice, 35
Woods Hole Oceanographic Institute, 118

Young ice, 35